STATISTICAL TOOLS OF SAFETY MANAGEMENT

Statistical Tools
of
Safety Management

Henry A. Walters

VAN NOSTRAND REINHOLD

I(T)P A Division of International Thomson Publishing Inc.

New York • Albany • Bonn • Boston • Detroit • London • Madrid • Melbourne
Mexico City • Paris • San Francisco • Singapore • Tokyo • Toronto

I⟨T⟩P™ A division of International Thomson Publishing, Inc.
The ITP Logo is a trademark under license

Printed in the United States of America

For more information, contact:

Van Nostrand Reinhold
115 Fifth Avenue
New York, NY 10003

Chapman & Hall GmbH
Pappelallee 3
69469 Weinheim
Germany

Chapman & Hall
2-6 Boundary Row
London
SE1 8HN
United Kingdom

International Thomson Publishing Asia
221 Henderson Road #05-10
Henderson Building
Singapore 0315

Thomas Nelson Australia
102 Dodds Street
South Melbourne, 3205
Victoria, Australia

International Thomson Publishing Japan
Hirakawacho Kyowa Building, 3F
2-2-1 Hirakawacho
Chiyoda-ku, 102 Tokyo
Japan

Nelson Canada
1120 Birchmount Road
Scarborough, Ontario
Canada M1K 5G4

International Thomson Editores
Campos Eliseos 385, Piso 7
Col. Polanco
11560 Mexico D.F. Mexico

1 2 3 4 5 6 7 8 9 10   QEB-FF 01 00 99 98 97 96 95

Library of Congress Cataloging-in-Publication Data

Walters, Henry A.
    Statistical tools of safety management/Henry A. Walters.
        p.   cm.
    Includes bibliographical references.
    ISBN 0-442-02125-9 (hardcover)
    1. Industrial safety—Statistical methods. I. Title
T55.3.S72W35   1995
363.11'9—dc20   95-19886

Contents

Preface

The safety professional can no longer afford the luxury of reacting to accident investigations and providing only qualitative assessments of hazards and risks. He or she has the opportunity and the obligation to perform quantitative analyses on a variety of safety problems. The fields of system safety and process safety management require people who know how to use statistical and analytical techniques.

This guide was written with two purposes in mind. The first was to provide a study guide to safety students who may or may not possess a strong background in probability and statistics. Traditional safety texts often fail to sufficiently explain all the steps between presenting a problem and solving it. This guide is intended to bridge the gap between theory and practice, and is designed to be used with available texts on system safety. It contains less theoretical discussion of why certain concepts work, and provides more examples of how to use the concepts in the workplace.

The second purpose is to provide a refresher for those safety professionals who have not used statistical and analytical methods often enough to remain proficient in these areas. Many of the techniques discussed will prove useful on certification examinations, such as the Certified Safety Professional examination.

By reading the brief discussion of the concept and then working the examples, the reader should be able to rapidly understand how a solution is derived. The technique used is that of programmed text. This means the reader will be taking an active part in his or her learning. After each sample question, at least four answers are given. The reader should solve the problem and select the answer that is closest to his or her answer and should then turn to the page numbered in parentheses after the answer and find the corresponding answer numbered. The solution informs the reader if the answer is correct or incorrect and explains how it was derived. I caution the reader to actively seek a solution to each problem before turning to the answers. It is quite easy to think one understands a solution because one can follow the example. I have found that it is a lot harder to work a problem myself than to follow the example.

The figures and tables are almost exclusively sample tables and figures for the specific problem addressed. They would be almost meaningless outside of the context of the sample problem, and a separate index of figures and tables would serve little purpose.

It is my sincere desire that this guide aid safety professionals with their study of probability and statistics and help them integrate these methods in their professional duties.

Acknowlegments

This book is the result of my doctoral program.

So many people provided assistance during my program that I almost hesitate to name them for fear of leaving someone out. However, I would be remiss if I missed this opportunity to thank those who played the larger roles in this effort.

Without the individual dedication of each member of the committee, there is no way that I could have completed the program in the time frame and the manner in which I desired. As the Core, Dr. Fritz was precisely that. He kept me on track and provided the much-needed moral support. Both adjuncts, Dr. Colligan and Dr. Krause, proved invaluable in their suggestions as to how best to study the field of occupational safety and health management. For such busy people, they were amazingly available to provide assistance when needed. My peers played their role superbly. Gladys Hankins was both a subject expert and the friend I needed on the same journey. Ed Rosser was the "older" brother who guided me through the process. I am looking forward to carrying these friendships into the future. Although one did not get the opportunity to really know the second reader, I feel as if Dr. Warner had nothing but my best interest at heart. Dr. Warner's contributions added breadth to the program that might not have otherwise been there, and her help in editing the final copy is immensely appreciated. Thank you, committee.

Without the support and patience of the faculty and staff in the Department of Occupational Safety and Health at Murray State University, my sanity would have left a long time ago. Dr. Friend and Dr. Kraemer worked with me to ensure that I had time to perform my Union study and still fulfill my obligations at the university. Ms. Loretta Powell and her student workers provided the administrative support, without which I could not have survived. Their willingness to do whatever I asked was beyond the call of duty. Their efforts are appreciated.

It's an old cliché, but, last but not least, I want to thank my family for putting up with me over the last two years. Much of the time that should have been spent with them was spent in the library, at seminars or classes, or hidden away with my computer. I want to thank my wife Terri and our daughters, Christie and Stephanie, for their support and recognition that education is important. It seems as if either Terri or I have been in school ever since we were married over 20 years ago. Maybe now we can have a little recess.

STATISTICAL TOOLS OF SAFETY MANAGEMENT

1

General Review of Mathematics Principles

A review of certain mathematical symbols, notations, and conventions may help you use this guide. This review is not all-inclusive, nor will it necessarily provide a discussion of the theory behind the concepts. It will, however, help you read and solve statistical formulae. The presentation in this chapter is in no particular order of importance or usefulness.

SIGNIFICANT DIGITS

The first convention is that of significant digits. The following basic rules determine whether a digit is significant.

All nonzero digits are significant.

No zero to the left of a nonzero digit is significant unless it is between two nonzero digits. For example, in the numbers 012 and .001 none of the zeros is significant. In the numbers 101 and 2003, all of the zeros are significant.

Zeros appearing to the right of significant digits may or may not be significant. Their significance will be determined by other factors. Zeros appearing to the right of significant digits but to the left of a decimal point are considered significant. A zero to the right of a significant digit to the right of a decimal point is considered significant. For example, in the numbers 1450. and .2360 the zeros are significant.

Additional rules govern significant numbers, but for our purposes here, the foregoing are all that you need. In this guide, all answers contain a minimum of four significant digits. When working problems, always carry your work to at least six significant digits. When a series of 9's occurs as the lead digits of a solution (i.e., .999994506), carry the answer to at least four significant digits after the last 9. The reason is that, in probabilities, you may be dealing with such high reliabilities that if you list an answer as .99999, without additional digits, it cannot be determined if you really have the correct solution. This convention violates the usual conventions of accuracy and precision, but it is necessary to ensure proper technique for solving problems in this guide.

Most calculators can be programmed to carry a specified number of digits. However, if necessary, you can always recall enough digits by using a simple trick. Say the calculator displays .000001, and you want the next four digits. First, write down the answer or store it in memory (for correct placement of the decimal point). Next, multiply the number displayed by 10,000, which moves the display four digits to the right and gives you the next four digits. You can work with your calculator to discover other methods of retrieving nondisplayed digits. Regardless of the display, the calculator is using the entire number. For example, if you divide 1 by 3, the calculator may only display .333333, but it is actually using that number as .33$\overline{3333}$, which means the 3's continue. This fact is important, because if you fail to use enough significant digits when solving a problem, the probability is high that your answer will contain considerable rounding error.

SCIENTIFIC NOTATION

The next convention is that of scientific notation, a method of taking numbers and standardizing their presentation. Numbers in scientific notation always consist of an integer to the left of the decimal point and the rest of the number after the decimal point. The entire number is then expressed as a multiple of 10. For example, the following numbers in the left column can be written as shown in the right column:

Number	Scientific notation
.01	$1.0 * 10^{-2}$
.1	$1.0 * 10^{-1}$
1	$1.0 * 10^{0}$
10	$1.0 * 10^{1}$
100	$1.0 * 10^{2}$

Note that the power of 10 is determined by the number of decimal places to the right (positive exponent) or left (negative exponent) of the first significant digit. Common practice and many calculators may show the above numbers as follows:

Number	Scientific notation
.01	1.0 E-2
.1	1.0 E-1
1	1.0 E0
10	1.0 E1
100	1.0 E2

Regardless of which notation is used, the sets of numbers are the same. Further examples of scientific notation include the following:

Number	Scientific notation
.001495	1.459 E-3
.01023	1.023 E-2
.2121	2.121 E-1
18	1.8 E1
215	2.15 E2

We stated that you should maintain at least six significant digits while solving problems. However, when the final solution is obtained, you may present the answer with only four significant digits (other than a string of 9's). Therefore, the rules for rounding must be discussed. This guide will use the engineering rules for rounding:

If the last digit is less than 5, round down (.12334 becomes .1233).
If the last digit is greater than 5, round up (.12336 becomes .1234).
If the last digit is 5, round to the nearest even number (.12335 becomes .1234 and
.12345 becomes .1234).

FACTORIALS

The next concept is that of factorials. A factorial is symbolized by an exclamation mark (!). This symbol means multiply the number by 1 less than the number, then by 2 less than the number, then by 3 less, and so on, until 1 is reached. For example,

$$9! = 9 * 8 * 7 * 6 * 5 * 4 * 3 * 2 * 1 = 362,880$$

Negative factorials or factorials containing decimals do not exist. All solutions are positive integers.

By definition, 0! equals 1, so you may divide by 0!. Thus, 10/0! = 10 and 2 * 0! = 2.

Fractions containing factorials *cannot* be reduced as with other numbers. For example,

$$4! \div 2! \neq 2!$$

because

$$(4 * 3 * 2 * 1) \div (2 * 1) = 12, \quad \text{but } 2! = 2 * 1 = 2$$

Likewise, factorials cannot be multiplied. For example,

$$3! * 2! \neq 6!$$

because

$$(3 * 2 * 1) * (2 * 1) = 12, \quad \text{but } 6! = 6 * 5 * 4 * 3 * 2 * 1 = 720$$

However, there is a way to reduce factorials. Let's assume you need to solve $100! \div (5! * 95!)$. Most calculators have a factorial function, but many do not perform calculations as high as 100!. Therefore, you must understand how reductions can be made. Like factorials can be reduced:

$$4! \div 4! = 1$$

where

$$(4 * 3 * 2 * 1) \div (4 * 3 * 2 * 1) = 1$$

How does that help solve $100! \div (5! * 95!)$? Well, 100! is simply $100 * 99 * 98 * \cdots * 1$, and 95! is $95 * 94 * 93 * \cdots * 1$. Thus, many terms repeat and can be canceled. Therefore,

$$\frac{100!}{5! * 95!} = \frac{100 * 99 * 98 * 97 * 96 * 95!}{5 * 4 * 3 * 2 * 1 * 95!} = \frac{100 * 99 * 98 * 97 * 96}{5 * 4 * 3 * 2 * 1}$$

$$= \frac{9,034,502,400}{120} = 75,287,520$$

Factorials are important in permutations and combinations. Permutations means the number of ways in which a series of events can occur, where the order of the series matters. For example, the effect produced by relay 1 closing before relay 2 closes is different from the effect produced if relay 2 closes before relay 1. The specific order of the number of ways that the relays could be open or closed is important. The general formula for permutations is:

$$\# \text{ permutations} = \frac{n!}{(n - x)!}$$

where n = number of objects from which to select
$\quad\quad x$ = number of places to put the objects

As another example of permutations, suppose you have 12 resistors and need to use 3 of them. Because of the resistors' properties, their actual circuit placement affects the use of the three resistors selected. In how many ways could these resistors be placed?

$$\frac{n!}{(n - x)!} = \frac{12!}{(12 - 3)} = \frac{479,001,600}{9!} = \frac{479,001,600}{362,880} = 1320$$

If, however, the order of a group of items does not affect the result, we are concerned with the combinations. The general formula for combinations is

$$\text{\# combinations} = \binom{n}{x} = \frac{n!}{x!(n-x)!}$$

where n = number of objects from which to select
x = number of places to place the objects

Substituting the numbers 12 and 3 from the resistor example gives

$$\frac{12!}{3!(12-3)!} = \frac{479,001,600}{3!9!} = \frac{479,001,600}{6 * 362,880} = 220$$

Permutations and combinations have different uses in safety applications, which are discussed later in the guide. Of importance now is to note that various textbooks will use different symbols for the combinations' formula. Regardless of the letters used within the ( ), the ( ) with two letters inside of them, one above the other without a ÷ sign, means to use the formula for combinations. For example

$$\binom{n}{x} = \frac{n!}{x!(n-x)!} = \binom{a}{b} = \frac{a!}{b!(a-b)!}$$

Many calculators perform permutations and combinations. It is beyond the scope of this guide to discuss how to use the calculator, but using a calculator can make solving permutations and combinations much easier.

How can you remember where to place the numbers in the combinations formula? Notice that, because the denominator is $x!\,(n-x)!$, after you subtract the x from the n in the numerator, the addition of the two remaining numbers must equal n. Therefore, if the sum of the two factorials in the denominator does not equal n, you made a mistake.

Some simple rules can reduce computation time. In the combinations formula

Anything over 0 equals 1:

$$\binom{n}{0} = \frac{n!}{0!(n-0)!} = \frac{n!}{0!n!} = 1$$

Anything over 1 equals itself:

$$\binom{n}{1} = \frac{n!}{1!(n-1)!} = \frac{n(n-1)!}{1!(n-1)!} = n$$

Anything over itself equals 1:

$$\binom{n}{n} = \frac{n!}{n!(n-n)!} = \frac{n!}{n!0!} = 1$$

Anything over itself minus 1 equals itself:

$$\binom{n}{n-1} = \frac{n!}{(n-1)!(n-(n-1))!} = \frac{n(n-1)!}{(n-1)!\ 1!} = n$$

With these shortcuts and conventions, you can examine some practical applications of probability from a safety perspective, but first try the sample problems.

SAMPLE PROBLEMS

1.1. The number 4,608 can be rewritten as

   a. 4.608 E-3   (230)
   b. 46.08 E2   (234)
   c. 4.608 E3   (259)
   d. 46.08 E-2   (236)

1.2. The number .0005798 can be rewritten as

   a. 5.798 E4   (177)
   b. 5.798 E3   (192)
   c. 5.798 E-3   (237)
   d. 5.798 E-4   (254)

1.3. The number 1,234,879 can be rewritten as

   a. 1.234 E6   (190)
   b. 1.234 E7   (270)
   c. 1.235 E6   (174)
   d. 1.235 E-6   (241)

1.4. The number .0286943 can be rewritten as

   a. 2.870 E-2   (258)
   b. 2.869 10-2   (264)
   c. 2.869 E2   (254)
   d. 2.870 10-2   (244)

1.5. The number 24895 can be rewritten as

   a. $2.489\ 10^4$ (169)
   b. 2.490 E-4   (247)
   c. $2.490\ 10^{-3}$ (241)
   d. 2.490 E4   (250)

1.6. $5! =$

    a. 120   (234)
    b. 115   (251)
    c. 15   (192)
    d. 720   (173)

1.7. $8!/4! =$

    a. $2!$ (229)
    b. 1680   (254)
    c. 24   (195)
    d. 4   (178)

1.8. $3! * 3! =$

    a. 81   (255)
    b. $9!$ (207)
    c. 1   (257)
    d. 36   (190)

1.9. $\dbinom{100}{10} =$

    a. 10   (209)
    b. 1.731 E13   (165)
    c. 1.902 E11   (168)
    d. 6.282 E19   (171)

1.10. If order matters, in how many ways can 4 parts be selected from a group of 20 parts?

    a. 80   (174)
    b. 5   (212)
    c. 4845   (246)
    d. 116,280   (177)

1.11. If order does not matter, in how many ways can 3 parts be selected from a group of 27 parts?

    a. 9   (199)
    b. 2925   (271)
    c. 17,550   (180)
    d. 362,880   (248)

1.12. $\dbinom{1000}{0} =$

    a. 1   (183)
    b. 10   (235)
    c. 100   (190)
    d. 1000   (242)

1.13. $\begin{pmatrix} 1000 \\ 5 \end{pmatrix} =$

    a. 295   (185)

    b. 9.9003495 E14   (279)

    c. 8.2502913 E12   (216)

    d. 200   (188)

2

General Probability

The probability of anything occurring can be defined as the number of ways it can occur divided by the number of exposures in which it could occur. That is

$$P = \text{occurrences} \div \text{possibilities}$$

There are two primary methods of determining probabilities: (1) *A priori* probabilities are those that can be determined before the fact, based on the inherent nature of the events. For example, a coin has only two sides. If one side is called heads and the other tails, the probability of heads in one flip of the coin is

$$P(\text{heads}) = 1 \text{ (one head)} \div 2 \text{ (two sides)} = 1/2 \text{ or } .5$$

It does not matter how many heads have been flipped in a row, the probability of heads on the next toss is still 1/2 (.5). A more dynamic example might be that of an honest die. The probability of rolling a 3 in one roll of a die is

$$P(1) = 1 \text{ (only one 3)} \div 6 \text{ (six numbers on a die)} = 1/6 \text{ or } .1666$$

The same is true for rolling any single number in one roll.

Unfortunately, there are not very many useful applications of a priori probability in safety. The best example may be switch positions. Without considering complications such as ergonomic design, we could say that a two-position switch has a probability of .5 of being "on" and a probability of .5 of being "off." A three-position switch has a probability of .333 of being in any one position, and so forth. There are few safety applications of a priori probability.

(2) *A posteriori* probabilities are those that cannot be determined before the fact. For example, if you want to know the probability of having an accident while traveling home, you know that you can have only one accident (barring rare occurrences which will be discussed later), but you do not know all of the various ways that the accident might occur (exposures). Therefore, you want to determine a best estimate probability based on past experience. In other words, you are going to base your probability of a future accident on what you have experienced in the past. Herein lies a small problem. Let's walk through a scenario to demonstrate the problem.

If you made seven trips home in the last week without an accident, you might think that the probability of an accident on the way home is 0/7 or 0, but is this realistic? Not really. There has not been enough history upon which to base your probability of an accident. Now, suppose that the previous week you had three accidents during your seven trips home. If we add this history to our data, we have a probability of

$$P(\text{acc}) = 3 \ (\text{no. acc}) \div 14 \ (\text{trips home})$$
$$= 3/14 \text{ or } .2143$$

This number means that you could expect to have one accident every four or five trips to your home. Is this realistic? Perhaps. But suppose those three accidents had been the only accidents you ever had? Since you have driven home every day for, say, eight years, would not a better probability be

$$P(\text{acc}) = 3 \ (\text{acc}) \div (8 \text{ yr} \times 365 \text{ days/yr})$$
$$= 3/2920 \text{ or } .001027$$

Again the answer is perhaps. If you never exceeded the speed limit until the week of your accidents, but you plan to speed from now on, and speed was a factor in your accidents, then .001027 is likely not a good estimated probability. Herein lies the problem. Of the three probabilities calculated, which one should you use to estimate the probability of having an accident on your way home today?

Safety professionals must make such determinations. The answer here, possibly, is that none of the probabilities should be used, simply because many factors determine whether an accident will occur. You may not have been at fault in any of the three accidents, or your speed may have caused them all, so you have decided never to speed again. Therefore, a better probability to use might be based on the experience of a larger sample. In this case, you might use a probability based on the experience of the United States' population as a whole, which could be obtained from an organization such as the American Automobile Association or the National Safety Council.

In a work situation you might want to use a probability based on the experience of the industry in which you work. Suppose you want to know the probability that your relatively new taxi company will have an accident on any given trip. The company does not yet have any real data on which to base your estimate. However, you could get data from another company and base your probability estimate on that data. For example, in the last 10 years ACME Taxi has the following experience:

Year	Accidents	Trips
1983	8	1000
1984	10	1250
1985	6	1100
1986	9	1150
1987	12	1300
1988	3	1350
1989	4	1400
1990	2	1500
1991	3	1475
1992	2	1600

You might want to use all of the data to determine the probability of an accident. The calculation is

$$P(\text{acc}) = 59 \text{ (total acc)} \div 13125 \text{ (total trips)}$$
$$= 59/13125 = .004495$$

However, a closer examination of the data is warranted. Between 1987 and 1988 something appears to have affected the accident rate. Perhaps taxis were made safer, or legislation was passed affecting safety, or the company instituted a new safety program. Whatever the reason, the rate of accidents was reduced. You should examine the reason for the change and see if it applies to your situation. After such examination, you discovered that in 1987 a law was passed requiring certification of taxi drivers. Since your company has to certify its drivers, which appears to have made a difference in ACME's experience, you might not want to use any data prior to 1988. Thus, the new probability is

$$P(\text{acc}) = 14 \text{ (acc between 88 and 92)} \div 7325 \text{ (trips between 88 and 92)}$$
$$= 14/7325 = .001911$$

In fact, other factors could have contributed to the lower numbers starting in 1990. If these apply, then the best estimated probability of an accident might be

$$P(\text{acc}) = 7 \text{ (acc 90–92)} \div 4575 \text{ (trips 90–92)}$$
$$= 7/4575 = .001530$$

However, you must be careful not to bias the probability by choosing which data to use. Since statistics are based on infinite occurrences, you should try to use as much data as possible. Generally, three to five years of data are sufficient. If so much data is not available, then the estimated probabilities can be generated with less data. You must try to get the best sample possible, based on similarities to the situation.

ADDITION LAW

Now that you know how to compute a simple probability, let's examine more interesting problems. First, remember that probability is always bound by 0 and 1. Thus, there is no such thing as a negative probability, and the highest probability possible is 1. This makes sense when you realize that the probability of rolling a 0 on a legal die is 0 (0 (number of zeros) ÷ 6 (total numbers)), and the probability of flipping a head or a tail in one flip of a coin must be 1 because there is nothing else that can logically happen (yes it could end up on its side, but that is so unlikely as to be excluded from probability). This leads to some interesting possibilities.

First, let's look at a simple example of the total probability equaling 1. A die has six numbers, and the probability of rolling each number is as follows:

Number	1	2	3	4	5	6
Probability	1/6	1/6	1/6	1/6	1/6	1/6

The probability of rolling a 1, a 2, a 3, a 4, a 5, or a 6 must be 1:

Number	1	or	2	or	3	or	4	or	5	or	6
Probability	1/6	+	1/6	+	1/6	+	1/6	+	1/6	+	1/6 = 1

This reasoning leads to the Addition Law, which means that if events are mutually exclusive (i.e., they cannot occur simultaneously) and you want the probability of something *or* something else occurring on one exposure, you simply add the individual probabilities. For example, the probability of rolling a 3 *or* a 5 in one roll is

$$P(3 \text{ or } 5) = 2/6 \text{ or } .3333$$

Number	3	or	5
Probability	1/6	+	1/6 = 2/6

The probability of rolling a 4 or a 5 or a 6 is

$$P(4 \text{ or } 5 \text{ or } 6) = 3/6 \text{ or } .500$$

Number	4	or	5	or	6
Probability	1/6	+	1/6	+	1/6 = 3/6 = .5

To give a safety example, assume that a relay can fail only in one of two ways. It can either fail closed or fail open, with the probability of failing closed being .05 and the probability of failing open being .1. Hence, the probability of the relay failing is

Failed position	open	or	closed
Probability	.05	+	.1 = .15

MULTIPLICATION LAW

Now suppose that we are going to roll a die twice. The probability of rolling a 3 on the first roll *and* a 4 on the second roll can be determined by the Multiplication Law: If two events are independent (i.e., one has no effect on the other), the probability of both events occurring in order is determined by multiplying the probabilities of each event occurring.

$$\begin{array}{llll}
\text{Number} & 3 \;\textit{and}\; 4 \\
\text{Probability} & 1/6 & * & 1/6 & = 1/36 \text{ or } .02778
\end{array}$$

It does not matter how many events occur. For example, the probability of rolling a 3 and a 5 and a 6, in that order, in three rolls of the die is

$$P(3 \text{ and } 5 \text{ and } 1) = 1/6 * 1/6 * 1/6 = 1/216$$

If you roll three dice at the same time, the probability is the same as rolling one die three times, as long as you kept track of which die was which. If you fail to identify which die is which, a combination of the Addition and Multiplication Laws applies.

If order did not matter in the first example of the Multiplication Law, then the Addition Law also applies because you want the probability of a 3 *and* a 4 *or* a 4 *and* a 3. First, you find the probability of rolling a 3 and a 4:

$$\begin{array}{lllll}
\text{Number} & 3 & \text{and} & 4 \\
\text{Probability} & 1/6 & * & 1/6 = 1/36
\end{array}$$

Then, find the probability of rolling a 4 *and* a 3:

$$\begin{array}{lllll}
\text{Number} & 4 & \text{and} & 3 \\
\text{Probability} & 1/6 & * & 1/6 = 1/36
\end{array}$$

Then the probability of rolling a 3 and a 4 *or* a 4 and a 3 is

$$1/36 + 1/36 = 2/36 \text{ or } .05556$$

A table of probabilities of rolling a pair of dice might help. Table 2-1 lists all of the ways in which a pair of dice can be rolled. The probability of rolling a specific number is in the right column.

Computing the probability of rolling the possibilities in the table is easy. But if you have a die with 100 sides, and each side is numbered, what is the probability of rolling higher than a 3? It is

$$\begin{array}{llllllllllll}
\text{Number} & 4 & \text{or} & 5 & \text{or} & 6 & \text{or} & 7 & \text{or} & 8 & \text{or} \cdots & 100 \\
\text{Probability} & 1/100 & + & 1/100 & + & 1/100 & + & 1/100 & + & 1/100 & + \cdots & 1/100 = 97/100
\end{array}$$

TABLE 2-1. PROBABILITIES FROM ROLLING A PAIR OF DICE

#	or	or	or	or	or		P(#)
1	—					= 0/36	= 0
2	1,1					= 1/36	= 1/36
3	1,2	2,1				= 1/36 + 1/36	= 2/36
4	1,3	2,2	3,1			= 1/36 + 1/36 + 1/36	= 3/36
5	1,4	2,3	3,2	4,1		= 1/36 + 1/36 + 1/36 + 1/36	= 4/36
6	1,5	2,4	3,3	4,2	5,1	= 1/36 + 1/36 + 1/36 + 1/36 + 1/36	= 5/36
7	1,6	2,5	3,4	4,3	5,2 6,1	= 1/36 + 1/36 + 1/36 + 1/36 + 1/36 + 1/36	= 6/36
8	2,6	3,5	4,4	5,3	6,2	= 1/36 + 1/36 + 1/36 + 1/36 + 1/36	= 5/36
9	3,6	4,5	5,4	6,3		= 1/36 + 1/36 + 1/36 + 1/36	= 4/36
10	4,6	5,5	6,4			= 1/36 + 1/36 + 1/36	= 3/36
11	5,6	6,5				= 1/36 + 1/36	= 2/36
12	6,6					= 1/36	= 1/36
13	—					= 0/36	= 0

However, since everything that can possibly be rolled must add to 1, there is a simpler way to determine this probability. It is called the Complementary Law.

COMPLEMENTARY LAW

Since all probabilities must sum to 1, for the example of a die with 100 sides the following equation must be true:

$$P(1,2,3,\cdots,100) = \underset{1/100}{1} + \underset{1/100}{2} + \underset{1/100}{3} + \cdots + \underset{1/100}{100} = 1.00$$

Therefore, the probability of rolling more than a 3 is found by subtracting the probability of rolling a 3 or less from 1. That is,

$$P(>3) = 1 - P(1,2,3) = 1 - (1/100 + 1/100 + 1/100)$$
$$= 1 - 3/100 = 97/100 \text{ or } .97$$

Thus, the Complementary Law simply means that if everything we don't want is subtracted from 1, the remainder is what we do want.

The Complementary Law is very useful in safety. The probability of any accidents can always be solved by finding the probability of no accidents (0) and subtracting it from 1:

$$P(\text{any acc}) + P(0) = 1$$
$$P(\text{any acc}) = 1 - P(0)$$

Let's try a safety example. Assume that the following probabilities are known. (Later you will learn how these probabilities may have been computed.)

Accidents	0	or	1	or	2	or	3	or	$\cdots$	or	400
$P(0,1,2,3, \cdots, 400) =$	.349	+	.387	+	.172	+	.081	+	$\cdots$	+	.000 = 1

The numbers represent the probability that a company would experience that exact number of accidents in 400 trips. In this example we assume that once an accident occurs, the trip is over. Therefore, in 400 trips there can be only 400 accidents. The total of all the probabilities of accidents the company might experience must equal 1.

Now assume we want to know the probability that the company will have less than three accidents. To ensure that you are solving for the correct answer, a drawing may assist. Simply write $P(r)$ from $P(0)$ to $P(r_{\text{int}} +1)$. Then use dots to indicate everything between $P(r_{\text{int}} +1)$ and $P(n)$. Next, draw a vertical line above the point of interest. For less than 3 it would appear as:

$$P(0) + P(1) + P(2) + P(3) + P(4) + \cdots + P(400) = 1$$

Notice that the line is between 2 and 3 since you do not want to include 3. Now which direction are we predicting? Draw an arrow in that direction. This appears as:

$$P(0) + P(1) + P(2) + P(3) + P(4) + \cdots + P(400) = 1$$

Therefore, the easiest way to solve this problem is to compute $P(0) + P(1) + P(2)$. Thus,

$$P(<3) = P(0) + P(1) + P(2)$$

If we had to determine the probability of three or fewer accidents, the picture would look like:

$$P(0) + P(1) + P(2) + P(3) + P(4) + \cdots + P(400) = 1$$

and be computed as:

$$P(\leq3) = P(0) + P(1) + P(2) + P(3)$$

Similarly, the probability of more than three accidents is

$$P(0) + P(1) + P(2) + P(3) + P(4) + \cdots + P(400) = 1$$

We could add $P(4) + P(5) + P(6) + \cdots + P(400)$, but that is tedious, so we use the Complementary Law:

$$P(>3) = P(4) + P(5) + P(6) + \cdots + P(400) = 1 - (P(0) + P(1) + P(2) + P(3))$$

Solving for $P(0) + P(1) + P(2) + P(3)$ is much easier than solving for $P(4) + P(5) + P(6) + \cdots + P(400)$.

DEPENDENCE VERSUS INDEPENDENCE

An additional factor of concern is dependence. Dependence is discussed in more detail in Chapter 11, but for our purposes here dependence simply means that what happens on one trial affects the probability of the next trial. For safety purposes, a parts bin is a good way to think of dependence. If 98 good parts are in a bin of 100 parts, then on the first draw from the bin, there is a probability of 98/100 of selecting a good part and a probability of 2/100 of getting a bad one. On the first draw, assume we get a good one. What is the probability of getting a bad one on the second draw?

Not so fast—this is a trick question. The answer depends on whether we are assuming independence. If no one replaces the part you drew out, then the probability of getting a bad part on the second draw is

2 (bad parts) ÷ 99 (total parts) = 2/99

The probability of getting a good part on the second draw is

97 (good parts) ÷ 99 (total parts) = 97/99

This problem exemplifies dependence—what occurs on one trial affects the next trial. When dealing with parts bins or the selection of almost anything from a finite group, you are relatively safe in assuming that the probabilities are dependent. However, to determine any given probability, you must have all of the needed information. In the previous examples then, you must have the total number of parts as well as the total of good parts or the total bad parts. If either of the two required numbers is missing, you cannot guess. For example, if you knew that there were two bad resistors in the bin, but you didn't know the total number of resistors, or the number of good ones, then you cannot assume the total number of resistors. If these quantities are not known, then you must make one of two assumptions, each of which leads to independence (one selection has no affect on the probability of another selection).

The first assumption is this: as fast as you take out a part someone returns the same kind of part to the bin. Thus, on the second draw the probability of getting a bad part is still 2/100 because 100 parts are still in the bin. What occurred on the first draw did not affect the probability of what occurred on the second draw.

The second assumption is this: the bin has so many parts that the selection of any part does not affect the next probability. For example, if 1 billion resistors were in the bin, is there really a difference between 2/1,000,000,000 and 2/999,999,999? Again, you could assume independence.

Be careful, though, about assuming independence too quickly because there will be significant differences in problem solutions of small quantities. For example, there is a significant difference between 1/100 and 1/99.

Let's now look at examples using the techniques we have learned. The first couple of sample problems will be of little practical value in real life, but they will help you understand the principles that will be useful in the examples that do relate to real-life situations.

SAMPLE PROBLEMS

2.1. A parts bin has 100 resistors: 40 are bad and 60 are good. On one draw, what is the probability of getting a bad resistor?

    a. .6000   (173)
    b. .0600   (190)
    c. .0400   (165)
    d. .4000   (266)

2.2. If you randomly pick a resistor out of the bin in Problem 2.1, someone replaces it with an identical resistor, then randomly draw another resistor, what is the probability of getting no bad resistors in both draws?

    a. 2.400 E-1   (175)
    b. 3.600 E-1   (194)
    c. .02400   (251)
    d. .03600   (180)

2.3. What is the probability of getting exactly one bad resistor in the two draws?

    a. 1.600 E-1   (271)
    b. .3600   (222)
    c. 4.800 E-1   (196)
    d. .2400   (275)

2.4. What is the probability of getting two bad resistors in the two draws?

    a. 3.600   (213)
    b. .1600   (250)
    c. 1.600 E-2   (241)
    d. .001600   (265)

2.5. What is the probability of getting any bad resistors in the two draws?

    a. .5200   (258)
    b. .8400   (200)
    c. .3600   (269)
    d. .6400   (203)

Solve the next five problems for the same bin of resistors, but *without replacement*. That is, as a resistor is drawn, it will *not* be replaced by another resistor.

2.6. What is the probability that in three draws three resistors are bad?

    a. .06110   (204)
    b. .06400   (268)
    c. .2160   (206)
    d. .2116   (227)

2.7. What is the probability that in three draws, you will draw a bad resistor, a good resistor, and a bad resistor in that order?

    a. .09600   (209)
    b. 9.600 E-1   (264)
    c. .09647   (273)
    d. 9.648 E-1   (265)

2.8. What is the probability that in three draws you will draw exactly two bad resistors?

    a. 9.6475 E-2   (201)
    b. .19295   (212)
    c. .28944   (215)
    d. 2.8942 E-1   (281)

2.9. What is the probability that in three draws you will draw two or more bad resistors?

    a. .3505   (249)
    b. .06110   (267)
    c. .06400   (219)
    d. .3520   (228)

2.10. What is the probability that in three draws you will draw any bad resistors?

    a. .21163   (203)
    b. .78400   (222)
    c. .21160   (277)
    d. .78837   (256)

PRACTICAL APPLICATIONS

Just as solving for

$$P(0) + P(1) + P(2) + P(3) + \cdots + P(n) = 1$$

proves tedious, solving dependent problems, such as the parts bin problems, can likewise be difficult if there are numerous draws from the bin. However, an equation has been developed for solving two-category (good/bad, cheap/inexpensive, black/white, etc.) dependent problems (r_{dep2}):

$$P(r_{\text{dep2}}) = \frac{\binom{n_1}{r_1}\binom{n_2}{r_2}}{\binom{n_t}{r_t}}$$

where n_1 = total of category 1 in lot
n_2 = total of category 2 in lot
n_t = total of lot
r_1 = number of category 1 being predicted
r_2 = number of category 2 being predicted
r_t = total needed

Remember that the parentheses mean use the formula for combinations in Chapter 1. The concept is that you need to determine the number of ways that category 1 can be selected and multiply that by the number of ways that category 2 can be selected. The total is then divided by the number of ways the needed objects can be selected from the total number of objects available.

There is a check, that if not met, will indicate that you have placed the numbers incorrectly in the formula. Note that the sum of the two n's in the numerator must equal the n in the denominator. This is logical since you need to account for all of the objects available. Also, the two r's in the numerator must equal the r in the denominator because you must account for all of the objects needed. An example might help.

Example

A parts bin of 100 resistors has 5 bad ones. What is the probability that a part containing 7 resistors will have exactly 1 bad resistor?

Solution   There are two categories: good or bad resistors. Taking parts from the bin leads us to believe that the probability is dependent (what happens on one draw affects the probability on the next draw). Hence,

$$P(r_{\text{dep2}}) = \frac{\binom{n_b}{r_b}\binom{n_g}{r_g}}{\binom{n_t}{r_t}}$$

where n_b = total bad resistors in lot
n_g = total good resistors in lot
n_t = total resistors in lot
r_b = number of bad resistors being predicted
r_g = number of good resistors being predicted
r_t = total resistors needed

Substituting the correct values into the formula yields

$$P(r_{\text{dep2}}) = \frac{\binom{5}{1}\binom{95}{6}}{\binom{100}{7}} = \frac{\left(\frac{5!}{1!\ 4!}\right)\left(\frac{95!}{6!\ 89!}\right)}{\frac{100!}{7!\ 93!}}$$

$$= \frac{\left(\frac{5\ *\ 4!}{1!\ 4!}\right)\left(\frac{95\ *\ 94\ *\ 93\ *\ 92\ *\ 91\ *\ 90\ *\ 89!}{6\ *\ 5\ *\ 4\ *\ 3\ *\ 2\ *\ 1\ *\ 89!}\right)}{\frac{100\ *\ 99\ *\ 98\ *\ 97\ *\ 96\ *\ 95\ *\ 94\ *\ 93!}{7\ *\ 6\ *\ 5\ *\ 4\ *\ 3\ *\ 2\ *\ 1\ *\ 93!}}$$

$$= \frac{\binom{5}{1}\left(\frac{625,757,605,200}{720}\right)}{\frac{8.0678106\ \text{E}13}{5,040}} = \frac{5\ *\ 869,170,785}{16,007,560,800}$$

$$= \frac{4,345,538,925}{16,007,560,800} = .271467901$$

Hence, there is a 27.15% probability that if 7 resistors were randomly selected from the batch of 100, 1 would be bad and 6 would be good. Note in the original information that the number of good resistors was not specified, but since there are 100 total resistors and you were told that 5 were bad, there must be 95 good ones. Likewise, the 6 good resistors needed were determined by the fact that you needed 7 total and only wanted 1 bad one. Also note that the check works ($1 + 6 = 7$ and $5 + 95 = 100$).

Example

Using the same information as the preceding example, find the probability of building a part with more than 1 bad resistor.

Solution   Do not simply subtract $P(1)$ from 1. That does not answer the question asked. Instead it determines the probability of not getting exactly 1 resistor. Using the previously discussed picture method

$$P(0) + P(1) + P(2) + \cdots + P(5) = 1$$

Note that since there are only 5 bad resistors, it is impossible to have more than 5 bad resistors, so the model stops at $P(5)$. We could sum $P(2)$ through $P(5)$, but using the Complementary Law is easier. This yields

$$P(>1) = 1 - (P(0) + P(1))$$

We must solve for $P(0)$ and $P(1)$. Since $P(1)$ was determined in the first example, we solve for $P(0)$.

$$P(0) \quad = \frac{\binom{5}{0}\binom{95}{7}}{\binom{100}{7}} = \frac{\left(\frac{5!}{0!\ 5!}\right)\left(\frac{95!}{7!\ 88!}\right)}{\frac{100!}{7!\ 93!}}$$

$$= \frac{(1)\dfrac{95 * 94 * 93 * 92 * 91 * 90 * 89 * 88!}{7 * 6 * 5 * 4 * 3 * 2 * 1 * 88!}}{\dfrac{100 * 99 * 98 * 97 * 96 * 95 * 94 * 93!}{7 * 6 * 5 * 4 * 3 * 2 * 1 * 93!}}$$

$$= \frac{(1)\dfrac{5.56924269\ \text{E}13}{5,040}}{\dfrac{8.0678106\ \text{E}13}{5,040}} = \frac{11,050,084,695}{16,007,560,800}$$

$$= .69030409024$$

$$P(<1) = 1 - \big(P(0) + P(1)\big)$$

$$= 1 - (.69030409024 + .271467901)$$

$$= 1 - .97177199124$$

$$P(>1) = .03822800876$$

Although we have solved these problems manually, it is much easier to use a calculator. Now see if you can do some problems on your own.

SAMPLE PROBLEMS

2.11.   In a batch of 200 smoke detectors, 95% of them are good. If you random-ly select 10 detectors to place in a building, what is the probability that the building will get any bad smoke detectors?

a. .6128		(279)
b. .5914		(226)
c. .3268		(227)
d. .4085		(230)

2.12.   If a parts bin contains 4 bad capacitors and 146 good ones, what is the probability that a part using 5 of these capacitors will be built with all good capacitors?

a. .1281		(259)
b. .8720		(262)
c. .1280		(217)
d. .8719		(209)

2.13. If a bin contains 500 hard hats, of which 5 are assumed to be defective, what is the probability that a work crew needing 12 hats will pick out 2 defective ones?

    a. 4.978 E-3   (265)
    b. 9.9502 E-1   (193)
    c. 1.148 E-1   (267)
    d. 8.852 E-1   (268)

2.14. A video tube containing 6 resistors will catch fire if 2 resistors are defective. If a bin contains 1000 resistors, of which 990 are good, what is the probability of building a tube with less than 3 bad resistors?

    a. .94134   (220)
    b. .05734   (203)
    c. 1.3085 E-3   (178)
    d. 9.9998578 E-1   (272)

2.15. For Problem 2.14 what is the probability of building a video tube that will catch fire?

    a. .99568   (275)
    b. .99869   (202)
    c. .001308   (278)
    d. .001323   (280)

3

Binomial Distribution

The perfect example of a binomial distribution is a coin toss. On any given toss, only one of two things can happen—a head or a tail. Also, the outcome of the first toss has no effect on the probability of the outcome of the second toss. This demonstrates the rules that apply to a binomial distribution. First, on any given trial, there can be only one of two outcomes, and the probability of each is known. Second, each trial is independent of the other. Because of these attributes, the mean (average) number of a binomial distribution is simply the number of trials multiplied by the individual probability per trial. Thus, if we are going to take 100 trips and the probability of an accident on any trip is .04, then the mean number of accidents in 100 trips is $100 * .04 = 4$. The importance of the mean will be demonstrated later.

From a safety perspective, one of the best examples of a binomial distribution is a transportation system. If we assume that we will count any trip, regardless of distance or time, as one exposure, then we might say that on any trip only two things can happen—we can have a safe trip or an accident. Furthermore, from previous experience (*a posteriori* probability), we can determine the probability of an accident on any trip. Knowing this, we can also determine the probability of a safe trip: $1 - P(\text{acc}) = P(\text{safe})$. The trips are independent of each other, so what happens on one trip does not affect the probability of an accident on the next trip. (True, arguments can be made against this such as learning curves, experience, etc., but for practical purposes the assumption of independence in this situation is legitimate.)

With this knowledge, can we now determine the probability of having one accident in two trips? Let's develop this with an example. Assume that the probabil-

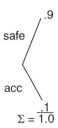

Figure 3-1

ity of an accident on any trip is .1. Hence, the probability of a safe trip is $1 - .1 =$ .9. Figure 3-1 demonstrates the possible outcomes of a single trip. Note that all that can happen equals a probability of 1.0.

Now let's make the second trip. Figure 3-2 diagrams that experience. If the first trip (upper branch) is safe, the second might be safe or an accident might occur. The same is true if an accident occurs on the first trip (lower branch). Note the corresponding probabilities to the various ways in which the two trips could occur. There is only one way to have no accidents and that is to have a safe trip followed by another safe trip ($P(0) = .81$).

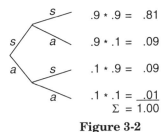

Figure 3-2

Likewise, there is only one way to have two accidents in two trips—an accident followed by another accident ($P(2) = .01$). However, there are two ways to have one accident in two trips. We could either have an accident followed by a safe trip, or we could have a safe trip followed by an accident. Since the probability of exactly one accident is .09, regardless of which way it occurs, the probability of having exactly one accident in two trips is $P(1) = .09 + .09 = .18$. Note that the probability of everything that can happen in two trips still equals 1:

$$P(0) + P(1) + P(2) = .81 + .18 + .01 = 1$$

What is the probability of at least one accident in two trips? There are several ways to derive the answer. However, the first step is to determine what answers the question. We use the method described in Chapter 2 and obtain Figure 3-3.

Having one accident or two accidents answers the question, we could simply add $P(1) + P(2)$: $P(\geq 1) = .18 + .01 = .19$. However, it is easier to simply subtract $P(0)$ from 1: $P(\geq 1) = 1 - P(0) = 1 - .81 = .19$.

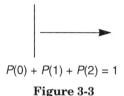

$$P(0) + P(1) + P(2) = 1$$

Figure 3-3

Now, if we made three trips, what is the probability of exactly two accidents? Figure 3-4 helps us solve this problem. Notice that there are three ways to experience exactly two accidents (*s,a,a* or *a,s,a* or *a,a,s*). Since they each have the same probability, we add their probabilities: $P(2) = .009 + .009 + .009 = .027$. This is quite easy, but consider the size of the diagram if there were 100 or 1000 trips? Thankfully, we don't have to solve it by hand.

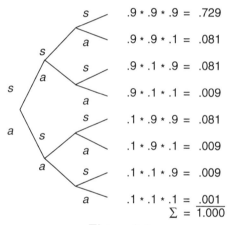

Figure 3-4

If on any exposure there is only one of two outcomes of known probability, and the exposures are independent of each other, then the binomial probability formula can be used to determine the probability of any mix of outcomes in any number of exposures. The formula is

$$P(r_{acc}) = \binom{n}{r} * P(acc)^r * (1 - P(acc))^{n-r}$$

where r = number of predicted accidents
  n = total number of exposures
  $P(\text{acc})$ = probability of accident on any exposure

The bold parentheses account for the number of combinations in which r accidents could occur in n exposures. Thus, it is merely a mathematical notation meaning to use the formula discussed in Chapter 1. Some texts use other letters within the ( ). It doesn't matter. The ( ) mean to use the combination's formula. To do this, you would simply substitute the same top letter into the same place in the formula (see the discussion in Chapter 1). For example:

$$\binom{n}{r} = \frac{n!}{r!(n-r)!} = \binom{b}{a} = \frac{b!}{a!(b-a)!}$$

The $P(\text{acc})^r$ accounts for the number of times that the probability of an accident must be multiplied. Remember that the Multiplication Law applies when you want to know the probability of an accident, an accident, . . . , and an accident. Thus the $P(1 - P(\text{acc}))^{n-r}$ accounts for the number of safe exposures that you want.

Here are some ways to help you substitute the correct numbers in the formula. The sum of the numbers in the denominator of the combination's formula must always equal the number in the numerator. The superscripts must equal n. Finally, the two probabilities must equal 1, since only two things can happen. These simple checks do not guarantee you will obtain a correct answer, but if any fail you are definitely wrong.

To solve a binomial probability, we must know n, r, and $P(\text{acc})$. They need not be provided explicitly, but you must be able to derive them from the given information. The following example might help you better understand the above discussion.

___

Example

In the last 100 flights of a particular type of single-engine aircraft, 2 engines failed. In the next 5 flights of this type of aircraft, what is the probability of no engine failures? What are the probabilities of 1, 2, 3, 4, and 5 engine failures?

Solution   The first question asks for $P(0)$. We know that

$$n \quad = \text{future exposures} = 5 \text{ flights}$$
$$r \quad = \text{predicted accidents} = 0$$
$$P(\text{acc}) = \text{occ/trials} = 2/100 = .02 \text{ (a posteriori prob)}$$

Substituting these values into the binomial formula yields

$$P(0) = \binom{5}{0} * (.02)^0 * (.98)^5$$

$$= \frac{5!}{0!(5-0)!} * (.02)^0 * (.98)^5$$

$$= \frac{5!}{0!\,5!} * (.02)^0 * (.98)^5$$

$$= 1 * 1 * .903920797 = .903920797$$

Note that the numbers in the denominator of the combination's formula equal 5, the number in the numerator, and the superscripts also equal 5. The probabilities for accidents and safe trips also equal 1.0.

Now let's turn to the next questions.

$$P(1) = \binom{5}{1} * (.02)^1 * (.98)^4$$

$$= \frac{5!}{1!\,4!} * (.02)^1 * (.98)^4$$

$$= 5 * .02 * .9223682 = .092236816$$

$$P(2) = \binom{5}{2} * (.02)^2 * (.98)^3$$

$$= \frac{5!}{2!\,3!} * (.02)^2 * (.98)^3$$

$$= \frac{5 * 4 * 3!}{2!\,3!} * (.02)^2 * (.98)^3$$

$$= \frac{20}{2} * .0004 * .9411920 = .003764768$$

$$P(3) = \binom{5}{3} * (.02)^3 * (.98)^2$$

$$= \frac{5!}{3!\,2!} * (.02)^3 * (.98)^2$$

$$= \frac{5 * 4 * 3!}{3!\,2!} * (.02)^3 * (.98)^2$$

$$= \frac{20}{2} * 8\ E\text{-}6 * .9604000 = .000076832$$

$$P(4) = \binom{5}{4} * (.02)^4 * (.98)^1$$

$$= \frac{5!}{4!\,1!} * (.02)^4 * (.98)^1$$

$$= 5 * 1.6\ E\text{-}7 * .98 = .0000007840$$

$$P(5) = \binom{5}{5} * (.02)^5 * (.98)^0$$

$$= \frac{5!}{5!\,0!} * (.02)^5 * (.98)^0$$

$$= 1 * 3.2\ E\text{-}9 * 1 = .000000003200$$

What is the probability of any engines failing? See Figure 3-5. Isn't there a possibility of only one engine failing? That doesn't answer the question because there is a separate and distinct possibility that two engines will fail. This logic

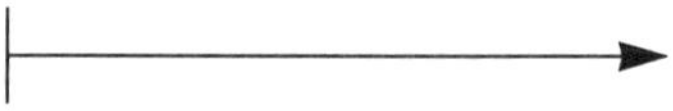

$$P(0) + P(1) + P(2) + P(3) + P(4) + P(5) = 1$$

Figure 3-5

continues up to the failure of five engines. (Since there is only one engine on each aircraft and only five flights of the aircraft, the probability is 0 of six engines failing.) To determine the solution you could sum $P(1)$ through $P(5)$, but using the Complementary Law is easier since there is only one way that there will be no engine failures ($P(0)$). Remember, the probability of any failures is always $1 - P(0)$. Thus,

$$P(\text{any}) = 1 - .903920797 = .096079203$$

What is the probability of less than two engines failing? See Figure 3-6.

$$P(0) + P(1) + P(2) + P(3) + P(4) + P(5) = 1$$

Figure 3-6

Simply add $P(0)$ and $P(1)$ because they are the only ways that less than two engines fail:

$$P(<2) = P(0) + P(1) = .903920797 + .092236816 = .996157613$$

What is the probability of more than one engine failing? According to Figure 3-7, we can use the solution of the last problem and subtract it from 1:

$$P(>1) = 1 - (P(0) + P(1)) = 1 - .996157613 = .003842387$$

$$P(0) + P(1) + P(2) + P(3) + P(4) + P(5) = 1$$

Figure 3-7

What is the probability of either three or four engines failing?

$$P(0) + P(1) + P(2) + P(3) + P(4) + P(5) = 1$$

$P(3 \text{ or } 4)$ $= P(3) + P(4) = .000076832 + .000000784 = .000077616$

Example

The probability of an accident for a trip is .03. What is the probability that in the next 100 trips more than 5 accidents will occur?

Solution   The easiest solution is the following:

$$P(0) + P(1) + P(2) + P(3) + P(4) + P(5) + P(6) + \ldots + P(100) = 1$$

$$P(> 5) = 1 - (P(0) + P(1) + P(2) + P(3) + P(4) + P(5))$$

$$P(0) = \binom{100}{0} * (.03)^0 * (.97)^{100} = \frac{100!}{0!\,100!} * (.03)^0 * (.97)^{100}$$

$$= 1 * 1 * .0475525 = .047552508$$

$$P(1) = \binom{100}{1} * (.03)^1 * (.97)^{99} = \frac{100!}{1!\,99!} * (.03)^1 * (.97)^{99}$$

$$= 100 * .03 * .0490232 = .147069612$$

$$P(2) = \binom{100}{2} * (.03)^2 * (.97)^{98} = \frac{100!}{2!\,98!} * (.03)^2 * (.97)^{98}$$

$$= \frac{100 * 99 * 98!}{2!\,98!} * (.03)^2 * (.97)^{98}$$

$$= \frac{9900}{2} * .0009 * .0505394 = .225152963$$

$$P(3) = \binom{100}{3} * (.03)^3 * (.97)^{97}$$

$$= \frac{100!}{3!\,97!} * (.03)^3 * (.97)^{97}$$

$$= \frac{100 * 99 * 98 * 97!}{3!\,97!} * (.03)^3 * (.97)^{97}$$

$$= \frac{970,200}{6} * .0000270 * .0521025 = .227474127$$

$$P(4) = \binom{100}{4} * (.03)^4 * (.97)^{96}$$

$$= \frac{100!}{4!\,96!} * (.03)^4 * (.97)^{96}$$

$$= \frac{100 * 99 * 98 * 97 * 96!}{4!\,96!} * (.03)^4 * (.97)^{96}$$

$$= \frac{94,109,400}{24} * .00000081 * .0537139 = .170605596$$

$$P(5) = \binom{100}{5} * (.03)^5 * (.97)^{95}$$

$$= \frac{100!}{5!\,95!} * (.03)^5 * (.97)^{95}$$

$$= \frac{100 * 99 * 98 * 97 * 96 * 95!}{5!\,95!} * (.03)^5 * (.97)^{95}$$

$$= \frac{9{,}034{,}502{,}400}{120} * 2.43 \text{ E-8} * .0553751 = .101308065$$

$$P(>5) = 1 - (P(0) + P(1) + P(2) + P(3) + P(4) + P(5))$$

$$= 1 - (.0475525 + .1470696 + .2251530 + .2274741 + .1706056 + .1013081)$$

$$= 1 - .9191629 = .0808371$$

Remember that the mean value is $nP(\text{acc})$. In the example notice that the individual probabilities between $P(0)$ and $P(5)$ rise between $P(0)$ and $P(3)$ (mean value) and then decrease. The values continue to decrease to $P(100)$. Therefore, for a problem with many values, be sure the values increase continuously to the mean and then continuously decrease. Once established there will never be a switch in direction. In the first example the mean was .01, so the values started decreasing immediately at $P(0)$.

It is possible to work a binomial problem backwards in order to establish what you would have to reduce your accident rate to in order to meet an established goal.

Example

Assume you want to achieve a probability of .995 that you will have no accidents in 500 trips. What is the probability of one accident per trip (p) that would have to be achieved?

Solution

$$P(0) + P(1) + P(2) + P(3) + P(4) + P(5) + P(6) + \ldots + P(100) = 1$$

$$P(0) \qquad = \binom{500}{0} * (p)^0 * (1-p)^{500} = .995$$

$$1 * 1 * (1-p)^{500} = .995$$

$$500 \log(1-p) = \log .995$$

$$\log(1-p) = \frac{\log(.995)}{500}$$

$$\log(1-p) = \frac{-.00217692}{500}$$

$$\log(1-p) = -.00000435385$$

$$(1-p) = \text{antilog}\,(-.0000044) = .999989975$$

$$p = .00001002$$

Now you should be ready to work the following problems on your own.

SAMPLE PROBLEMS

3.1. In 2000 exposures to a certain process, your company has had 4 accidents. In the next 1000 exposures, what is the probability that the company will have exactly 2 accidents?

    a. .72906   (229)
    b. .72933   (198)
    c. .27094   (239)
    d. .27067   (244)

3.2. Given the same data as Problem 3.1, what is the probability that the company will have no accidents?

    a. 1.3506 E1   (254)
    b. 1.3506 E-1   (169)
    c. 2.7067 E-1   (179)
    d. 2.7094 E-1   (187)

3.3. Given the same data as Problem 3.1, what is the probability that the company will have any accidents?

    a. .86494   (194)
    b. .86483   (196)
    c. .86485   (199)
    d. .86493   (202)

3.4. Given the same data as Problem 3.1, what is the probability that the company will have less than 4 accidents?

    a. .09022   (207)
    b. .18063   (210)
    c. .81937   (213)
    d. .85730   (220)

3.5. A parts bin contains 400 parts of which .02 are defective. If a machine using 8 of these parts is built, what is the probability that the machine contains 7 good parts?

    a. 1.3890 E-1   (169)
    b. 1.4122 E-1   (282)
    c. 2.2346 E-1   (236)
    d. 2.3246 E-1   (242)

3.6. A parts bin contains .02 defective parts. If a machine using 8 of these parts is built, what is the probability that the machine contains 7 good parts?

    a. .14869   (224)
    b. .13982   (261)
    c. .13890   (268)
    d. .14075   (274)

4

Multinomial Distribution

Just as the binomial distribution, the multinomial distribution must be independent, and there can be only one specific outcome of any trial. However, unlike the binomial distribution, there are more than two types of events that can occur from any multinomial trial. In the binomial example, you could either have an accident or a safe trip. Now let's assume that an accident equates to a loss. From a safety perspective, it might be more useful to categorize losses as different types. A ranking similar to that of the government's MIL-STD-881B might prove useful. Therefore, we would have four categories of loss: catastrophic, major, minor, and negligible. Another ranking might be fatality, major loss, minor loss, and safe trip.

As in the binomial distribution, we could use diagrams to calculate probabilities for a multinomial distribution. Or we can use the multinomial probability formula:

$$P(r_1, r_2, r_3, \ldots, r_k) = \frac{n!}{r_1! r_2! r_3! \ldots r_k!} (p_1)^{r_1} * (p_2)^{r_2} * (p_3)^{r_3} * \ldots * (p_k)^{r_k}$$

where n = total number of exposures
$r_1 - r_k$ = predicted number of each respective category
$p_1 - p_k$ = probability of each category on any exposure
(Subscripts 1–k mean that you can have as many categories as you want. You must be careful to match r_1 with p_1, etc.)

From a safety perspective, there are basically three questions that can be asked when we use a multinomial distribution: (1) What is the probability of a specific mix of categories, say 1 fatality, 4 major accidents, and 10 minor accidents in 100 trips? (2) What is the probability of any accidents? (3) What is the probability of no accidents? (Note: In the last two questions, the problem has been reduced to only two categories.) Some examples may help visualize these questions.

Example

A company has experienced 3 fatal accidents, 6 major accidents, and 12 minor accidents in its last 1000 trips. In the next year the company expects to make 300 trips. What is the probability of exactly 1 fatal, 2 major, and 4 minor accidents?

Solution First determine the probability of each type of accident:

$$P(\text{fat}) \;=\; p \;=\; \text{occ} \div \text{exp} \;=\; 3/1000 \;=\; .003$$
$$P(\text{maj}) \;=\; p_2 \;=\; \qquad\;=\; 6/1000 \;=\; .006$$
$$P(\text{min}) \;=\; p_3 \;=\; \qquad\;=\; 12/1000 \;=\; .012$$

Note that the probabilities do not add to 1 ($p_1 + p_2 + p_3 = .003 + .006 + .012 = .021$); therefore, something must be missing. Is it not logical to assume some safe trips? That is the missing category.

$$P(\text{safe}) \;=\; p_4 = 1 - (p_1 + p_2 + p_3)$$
$$= 1 - (.003 + .006 + .012) = 1 - .021 = .979$$

and $r_1 - r_k$ are

$$r_1 \;=\; \text{predicted fatal} \;=\; 1$$
$$r_2 \;=\; \text{predicted major} \;=\; 2$$
$$r_3 \;=\; \text{predicted minor} \;=\; 4$$

Just as in the binomial distribution, you must account for all of the ways in which a trip can end. Therefore, the sum of the values in the denominator must equal the figure in the numerator. Since $n = 300$ (total trips) and $7 \neq 300$, an r must be missing. Since we have discovered that there had to be a p_4, there must be an r_4 equal to the number of safe trips: $300 - 7 = 293$. Substituting these values into the formula yields

$$P(1, 2, 4, 293) = \frac{300!}{1!\,2!\,4!\,293!} * (.003)^1 * (.006)^2 * (.012)^4 * (.979)^{293}$$

$$= \frac{2.0381034\ \text{E}17}{48} * (.003)^1 * (.006)^2 * (.012)^4 * (.979)^{293}$$

$$= 4.24604875\ \text{E}15 * .003 * 3.6\ \text{E-}5 * 2.0736\ \text{E-}8 * 1.99218056\ \text{E-}3$$

$$= .0189435956$$

Remember the rules of reduction for large factorials. We find 300! as follows:

$$\frac{300 * 299 * 298 * 297 * 296 * 295 * 294 * 293!}{1 * 2 * 1 * 4 * 3 * 2 * 1 * 293!}$$

Now, what is the probability of no accidents? Since no accidents obviously means no fatalities, no major accidents, and no minor accidents, substituting into the formula yields

$$P(0,0,0,300) = \frac{300!}{0!\,0!\,0!\,300!} * (.003)^0 * (.006)^0 * (.012)^0 * (.979)^{300}$$

$$= 1 * 1 * 1 * 1 * (.979)^{300} = .00171714726$$

Remember that the probability of no accidents reduces the outcomes to two categories—accident or safe. Although the solution is correct, it might be faster to use the binomial distribution formula:

$$P(0) = \binom{300}{0}(.021)^0 * (.979)^{300}$$

$$= 1 * 1 * .00171714726 = .00171714726$$

The probability of an accident (.021) can be derived by one of two methods. The first is to add the probabilities of each category of accidents $(p_1 + p_2 + p_3)$. However, the second way is easier: just compute $1 - p_4$.

Now, what is the probability of any accidents? Remembering that "any" is always equal to $1 - P(0)$, we write

$$P(\text{any}) = 1 - P(0,0,0,300) = 1 - .00171714726 = .998282853$$

Thus, the probability of any accidents can also be solved by the binomial distribution.

Caution: It is logical to assume a category of safe trips. However, you should not make up or assume any other type of category. Therefore, if the a posteriori probabilities do not equal 1 and you have a probability for safe trips, do not assume another category until more information is available as to what it might be.

If a problem appears to be dependent (parts in a bin, selection from a group of things), but there is missing information that prohibits solving the problem by a dependent formula, then the multinomial distribution formula may be used. This is based on the assumption that either there are so many items from which to select that the problem approaches independence, or that the items are replaced as rapidly as they are selected.

SAMPLE PROBLEMS

Some of these are not realistic because of the small number of exposures. Statistics are based on large populations, so you would rarely attempt to make predictions with so few exposures. The small n is for convenience and ease.

4.1. Your trucking company has experienced the following probabilities of accidents:

$$P(\text{fat}) = .001$$
$$P(\text{maj}) = .014$$
$$P(\text{min}) = .025$$
$$P(\text{safe}) = .960$$

In the next year you will make 50 trips. What is the probability of exactly 0 fatalities, 1 major accident, and 2 minor accidents?

a.	1.5106 E-1	(243)
b.	.075532	(248)
c.	7.553 E-1	(253)
d.	cannot solve	(257)

4.2. Assuming the same data as Problem 4.1, what is the probability of no accidents?

a.	6.0426 E-3	(165)
b.	2.4681 E-1	(176)
c.	1.2988 E-1	(181)
d.	cannot solve	(185)

4.3. In the last 5000 exposures, your company has had 2 fatalities, 5 major accidents, and 10 minor accidents. In the next 1000 exposures, what is the probability of 1 fatality, 2 major accidents, and 4 minor accidents?

a.	2.1295 E-1	(187)
b.	4.4364 E-3	(191)
c.	1.5624 E-2	(193)
d.	cannot solve	(197)

4.4 Your parts bin contains both partially defective and totally defective parts. If a component requiring 9 parts must be made from this bin, and the probabilities of parts are as follows, what is the probability of building a component with 2 totally defective parts and 3 partially defective parts?

$$P(\text{totally}) = .1$$
$$P(\text{partially}) = .2$$
$$P(\text{good}) = .7$$

a. 2.4202 E-2	(203)
b. 5.8085 E-1	(209)
c. 2.9432 E-3	(216)
d. cannot solve	(273)

5

Hypergeometric Distribution

The hypergeometric distribution is the formal name of the dependent formula we discussed in Chapter 2. The conditions to be met to use the formula are identical to those for the multinomial distribution, with one exception. The formula for the hypergeometric distribution is for dependent probabilities. (See Chapter 2 to review dependency.) The formula is

$$P(r_1, r_2, r_3, \ldots, r_k) = \frac{\binom{w_1}{r_1}\binom{w_2}{r_2}\binom{w_3}{r_3}\cdots\binom{w_k}{r_k}}{\binom{w}{r}}$$

where $w_1, w_2, w_3, \cdots, w_k$ = total of each category of interest
$r_1, r_2, r_3, \cdots, r_k$ = predicted amount of each category
w = total objects
r = total objects needed

Remember that $w_1 + w_2 + w_3 + w_k = w$ and $r_1 + r_2 + r_3 + r_k = r$. All values must be known.

A good example of a hypergeometric situation which is related to safety is quality control. If you are trying to predict the probability of making a product with a certain mix of parts, then the hypergeometric formula is probably best. Again, three questions can be asked from a safety perspective: (1) What is the probability of a specific mix of categories, say 5 defective, 4 partially defective, and 10 good parts in a component of 19 parts made from a bin of 100 parts? (2) What is the probability of any defective parts? (3) What is the probability of no defective parts. (*Note:* In the last two questions, the problem is reduced to only two categories.) Some examples may help visualize these questions.

___

Example

An electronically controlled fire suppression system requires 10 identical relays. If a parts bin contains 200 relays of which it is believed that 6 are totally defective and 8 are partially defective, what is the probability of building a system containing 1 totally defective relay and 2 partially defective ones?

Solution

$$P(1,2,7) = \frac{\binom{6}{1}\binom{8}{2}\binom{186}{7}}{\binom{200}{10}}$$

$$= \frac{\left(\frac{6!}{1!5!}\right)\left(\frac{8!}{2!6!}\right)\left(\frac{186!}{7!179!}\right)}{\frac{200!}{10!190!}}$$

$$= \frac{6 * \dfrac{8 * 7 * 6!}{2 * 6!} * \dfrac{186 * 185 * 184 * \ldots * 180 * 179!}{7 * 6 * 5 * 4 * 3 * 2 * 179!}}{\dfrac{200 * 199 * 198 * 197 * \ldots * 192 * 191 * 190!}{10 * 9 * 8 * 7 * 6 * 5 * 4 * 3 * 2 * 190!}}$$

$$= \frac{6 * \dfrac{56}{2} * \dfrac{6.8703056 \ \text{E15}}{5040}}{\dfrac{8.1470204 \ \text{E22}}{3,628,800}} = \frac{2.2901019 \ \text{E14}}{2.2451004 \ \text{E16}}$$

$$= .010200443$$

Note: Although neither the number of good parts in the bin nor the number of good parts needed was provided, they could be determined from the given information since the *w*'s must equal the total relays available, and the *r*'s must equal the total relays needed.

___

Example

What is the probability of building a system with no defective relays?

Solution Using the hypergeometric formula and substituting in the values yields

$$P(0,0,10) = \frac{\binom{6}{0}\binom{8}{0}\binom{186}{10}}{\binom{200}{10}}$$

$$= \frac{1 * 1 * \dfrac{186!}{10!\,176!}}{\dfrac{200!}{10!\,190!}}$$

$$= \frac{1 * 1 * \dfrac{186 * 185 * 184 * \ldots * 177 * 176!}{10 * 9 * 8 * \ldots * 3 * 2 * 176!}}{\dfrac{200 * 199 * 198 * 197 * \ldots * 192 * 191 * 190!}{10 * 9 * 8 * 7 * 6 * 5 * 4 * 3 * 2 * 190!}}$$

$$= \frac{1 * 1 * \dfrac{3.8745597\ E22}{3,628,800}}{\dfrac{8.1470204\ E22}{3,628,800}} = \frac{1.0677248\ E16}{2.2451004\ E16}$$

$$= .475579961$$

Just as in the multinomial distribution, the above solution could have been derived by assuming two categories—good and bad. Then

$$P(0,0,10) = \frac{\binom{14}{0}\binom{186}{10}}{\binom{200}{10}}$$

$$= \frac{1 * \dfrac{186!}{10!\,176!}}{\dfrac{200!}{10!\,190!}}$$

$$= \frac{1 * \dfrac{186 * 185 * 184 * \ldots * 177 * 176!}{10 * 9 * 8 * \ldots * 3 * 2 * 176!}}{\dfrac{200 * 199 * 198 * 197 * \ldots * 192 * 191 * 190!}{10 * 9 * 8 * 7 * 6 * 5 * 4 * 3 * 2 * 190!}}$$

$$= \frac{1 * \dfrac{3.8745597\ E22}{3,628,800}}{\dfrac{8.1470204\ E22}{3,628,800}} = \frac{1.0677248\ E16}{2.2451004\ E16}$$

$$= .475579961$$

Example

What is the probability of any defective relays?

Solution

$$P(\text{any}) = 1 - P(0,0,10) = 1 - .475579961 = .524420039$$

The formula for the hypergeometric distribution behaves very similarly to the earlier, two-category, dependent formula. As the total number of items in the sample gets larger, it more closely approximates an independent problem. Therefore, you must have a total available for the numerator; you cannot assume one. The only assumption you can make if the totals are not available is that there are so many that independency can be assumed or that items are being replaced as fast as they are depleted. See how well you do on the following sample problems.

SAMPLE PROBLEMS

5.1. A container holds 80 cartridges for air-purifying respirators: 6 are for dust, 10 are for mists and vapors, and the rest are for organic materials. If you randomly select 9 cartridges, what is the probability of getting exactly 2 for dust and 4 for mists?

a. 1.3583 E-8 (281)
b. 5.6594 E-4 (273)
c. 4.5791 E-7 (269)
d. 8.6306 E-3 (266)

5.2. For the same data as Problem 5.1, what is the probability of getting all organic material cartridges in 9 draws?

a. 1.6073 E-1 (229)
b. .13422 (225)
c. 1.4752 E-1 (223)
d. .11876 (218)

5.3. For the same data as Problem 5.1, what is the probability of selecting any dust or mist cartridges in 10 selections?

a. 9.08002465 E-1 (214)
b. 8.81239546 E-1 (211)
c. 9.19975346 E-2 (204)
d. 9.91445173 E-2 (258)

5.4. A bin contains 500 fire alarms of which .028 are partially defective and .012 are totally defective. If you choose 8 alarms, what is the probability that you will select 1 totally defective one and 2 partially defective ones?

    a. 2.2930989 E-15   (229)
    b. 5.9620572 E-15   (195)
    c. 1.23978071 E-3   (191)
    d. 4.76838734 E-4   (189)

5.5. Assume the same data as Problem 5.4, except that the total number of alarms is unknown. If you need 8 alarms, what is the probability that you will select 1 totally defective one and 2 partially defective ones?

    a. 1.8966528 E-1   (186)
    b. 8.128512 E-2   (182)
    c. 5.52313896 E-4   (178)
    d. 1.28873242 E-3   (171)

6

Poisson Distribution

Let's examine our original example of a binomial distribution a little more closely. Some assumptions we made were that all trips were counted the same, as one exposure, and that once an accident occurred the trial was over so that there was never a possibility of two accidents on one trip. Are these assumptions always realistic?

Pretend you are in charge of safety for a trucking company, and the company needs to make two trips. The first trip is within city limits. The other trip is from Nashville, Tennessee, to Los Angeles, California. If a breakdown occurs on the first trip, it might be logical to assume that that is the end of the trip for that truck because you might transfer the material to another truck and have the broken truck repaired. Now, let's assume the second truck broke down in Denver. Since you don't have any other trucks in Denver, it is likely that you would have it fixed and send it on its way again. But it might break down again, say in Salt Lake City. Fixed once more, could it not break down again at the California border? This could go on indefinitely until the truck reached Los Angeles. Even within Los Angeles, it might break down before it reached its destination. Therefore, mathematically, there could be an infinite number of breakdowns on the one trip.

For this reason, the binomial distribution may not be proper in this situation. Rather than count a trip as one trip, it might be better to determine exposure based on some number of hours, miles, or other unit of equal measure for all trucks. Also, instead of assuming that only one event can occur per exposure, it might be better to assume that there could be an infinite number of occurrences per exposure. This is the principle behind the Poisson distribution.

Like the binomial distribution, the Poisson distribution is still dichotomous—

that is, an occurrence can be only one of two outcomes. As safety examples they could be safe versus accident, work versus break, or operate versus fail, and so on. However, there is a major difference between the two distributions. Since the trials are considered continuous in the Poisson distribution, it is impossible to distinguish between one truck having five accidents on a trip and five trucks having one accident each. The importance of this difference will be disclosed later.

The general formula for the Poisson distribution is

$$P(r) = \frac{(\lambda t)^r / e^{-\lambda t}}{r!}$$

where t = exposure period in the same units used for λ
  r = number of predicted occurrences
  e = constant (inverse of natural logarithm = @2.718)
  λ = a posteriori probability of occurrence of an exposure

Here λ is equivalent to p_1 in the binomial formula. Recall that the a posteriori probability is determined by dividing the past occurrence by the number of historical exposures. In the Poisson distribution, it is typically expressed as failures ÷ time. For example, if 10,000 hours of test time for a part resulted in five failures, λ is

$$\lambda = 5 \text{ failures} \div 10,000 \text{ hours} = .0005$$

Example

Assume that a smoke detector has been tested for 1000 hours and experienced 10 failures during that time. What is the probability of exactly 2 failures in 500 hours of operation?

Solution Since $t = 500$, $\lambda = 10/1000 = .01$, $r = 2$, we have

$$P(2) = \frac{(.01 * 500)^2 e^{-.01 * 500}}{2!}$$
$$= \frac{(5)^2 e^{-5}}{2} = \frac{25 * .006737947}{2}$$
$$= \frac{.168448675}{2} = .0842243375$$

Note that λ was determined by 10 failures ÷ 1000 hours of test. Therefore, t must be in hours. An easy mistake is to forget to make the superscript for e negative. There is a quick check for this mistake. Every e raised to a negative power is a positive number less than 1, and every e raised to a positive number is a number greater than 1. By definition, $e^0 = 1$.

Your calculator should have an e^x function. If not, see if it has the natural log function, usually "ln x," and use inverse ln x for e^x. Your calculator manual will explain how to use these functions. Try to verify the following:

$$e^0 \quad = 1$$
$$e^1 \quad = 2.718$$
$$e^2 \quad = 7.389$$
$$e^{-1} \quad = .3679$$
$$e^{-2} \quad = .1353$$
$$e^{-.5} \quad = .6065$$
$$e^{-.24} = .7866$$

Example

For the information in the previous example, what is the probability of no failures in 500 hours of use?

Solution Since $P(0)$ means that $r = 0$, when the first term of the numerator is raised to the zero power it simply equals 1. Likewise, by definition, $r! = 0! = 1$. Therefore, for $P(0)$, the Poisson formula reduces to

$$P(0) = \frac{(\lambda t)^0 e^{-\lambda t}}{0!} = \frac{1 * e^{-\lambda t}}{1} = e^{-\lambda t}$$

The solution is simply

$$P(0) = e^{-.01*500} = e^{-5} = .006737947$$

Notice several things in this example. First, no matter what r is, as long as the original data is not changed $e^{-\lambda t}$ remains constant. Therefore, if you are solving a problem requiring several $P(r)$'s, there is no need to recompute $e^{-\lambda t}$. Also, regardless of r, as long as the original data does not change λt remains constant, so there is no need to recompute the first term of the numerator before you raise it to the new r power.

Another interesting point is that λt equals the mean expected value (5 in this example). Therefore, just as in the binomial distribution, $P(r)$ increases until it reaches λt and then continuously decreases. In this example $P(r)$ increases until $P(5)$ and then decreases. To demonstrate:

$$P(0) \ = e^{-.01*500} = e^{-5} = .006737947$$
$$P(1) \ = \frac{(.01 * 500)^1 e^{-.01*500}}{1!}$$

$$= \frac{(5)^1 e^{-5}}{1} = 5 * .006737947 = .033689735$$

$$P(2) = \frac{(5)^2 e^{-5}}{2!} = \frac{25 * .006737947}{2} = .0842243375$$

$$P(3) = \frac{(5)^3 e^{-5}}{3!} = \frac{125 * .006737947}{6} = .140373896$$

$$P(4) = \frac{(5)^4 e^{-5}}{4!} = \frac{625 * .006737947}{24} = .175467370$$

$$P(5) = \frac{(5)^5 e^{-5}}{5!} = \frac{3125 * .006737947}{120} = .175467370$$

$$P(6) = \frac{(5)^6 e^{-5}}{6!} = \frac{15,625 * .006737947}{720} = .146222808$$

$$P(7) = \frac{(5)^7 e^{-5}}{7!} = \frac{78,125 * .006737947}{5,040} = .104444863$$

If you must perform such a series of calculations, you can use a time-saving technique. Notice that each increase by 1 of r simply changes the exponent for λt by 1 and the factorial in the denominator by 1. Increasing the factorial by 1 means you are multiplying the previous denominator by the new r. Therefore, it is possible to use the answer obtained for one r to determine the probability of the next r. For example, to solve for $P(1)$, multiply $P(0)$ by λt and divide by 1. To solve for $P(2)$, multiply $P(1)$ by λt and divide by 2. In other words, to solve for any $P(r)$ in a series, multiply the preceding $P(r)$ by λt and divide by the new r. For our above example

$$P(0) = e^{-.01*500} = e^{-5} = .006737947$$

$$P(1) = \frac{.006737947 * 5}{1} = .033689735$$

$$P(2) = \frac{.033689735 * 5}{2} = .0842243375$$

$$P(3) = \frac{.0842243375 * 5}{3} = .140373896$$

$$P(4) = \frac{.140373896 * 5}{4} = .175467370$$

$$P(5) = \frac{.175467370 * 5}{5} = .175467370$$

$$P(6) = \frac{.175467370 * 5}{6} = .146222808$$

$$P(7) = \frac{.146222808 * 5}{7} = .104444863$$

All probabilities still must equal 1, in the Poisson distribution

$$P(0) + P(1) + P(2) + P(3) + P(4) + P(5) + \cdots + P(\infty) = 1$$

An interesting twist is that it is impossible to solve a problem which asks the probability of $P(>r)$ without using the Complementary Law because there is no end to the number of r's possible.

In industry, it is not likely that someone will walk up to you with the historical raw data for failures. What is more likely is that they will have a number for the mean time between failures (MTBF), which is failures ÷ time, or the reciprocal of λ. Thus, if λ = occurrences/exposures and m = exposures/occurrences, then λ = $1/m$ or $m = 1/\lambda$.

Therefore, instead of changing MTBF to λ to solve the probability of a Poisson distribution, we can write the formula as

$$P(r) = \frac{(\lambda t)^r e^{-\lambda t}}{r!} = \frac{(t/m)^r e^{-t/m}}{r!}$$

where m = MTBF = time/failures
  t = exposure period in the same units used for m
  r = number of predicted occurrences
  e = constant (inverse of natural log = @2.7183)

To use the MTBF in the original example, only the initial form of the equation would change.

Example

A smoke detector has been tested for 1000 hours and experienced 10 failures during that time. What is the probability of exactly 2 failures in 500 hours of operation?

Solution

$$t \quad = 500 \quad m = 1000/10 = 100 \quad r = 2$$
$$P(2) = \frac{(500/100)^2 e^{-500/100}}{2!}$$
$$= (5)^2 e^{-5} = \frac{25 * .006737947}{2}$$
$$= \frac{.168448675}{2} = .084224337$$

Although the basic principle is true for all probabilities, another factor must be considered for the Poisson distribution. In almost every distribution, you should not attempt to predict an outcome for a number of exposures that is greater than the history on which the a posteriori probability is based. For example, in a binomial distribution problem, if you had only 500 trips, during which five accidents occurred, you should not attempt to predict the probability of r accidents in the next 1000 trips based on the a posteriori probability of .01 (5/500). This fact is especially important for the Poisson distribution. Remember that the distribution cannot tell if one product had four fires or if four products had one fire each. Therefore, the following example could lead to faulty conclusions.

Example

Ten parts were each tested for 15 hours and experienced a total of 3 failures. What is the probability that one part will experience no failures in 100 hours of operation?

Solution   At first glance this appears to be a regular Poisson distribution. We have tested the part for 150 hours, so $\lambda = 3/150 = .02$. The 150 hours of test is greater than the number of hours we are going to use the part. However, there is a problem. Each part in the test was tested for only 15 hours. Even though there have been more hours of test than we are trying to predict for the occurrence of r, we should not trust the prediction. Maybe at 20 hours of operation each part would self-destruct. Since we have not tested a single part for more than 15 hours, we should not try to predict a probability of a single part's lasting more than 15 hours. Although this principle is often violated, it should be adhered to.

As for the binomial distribution, we can set a standard for the probability of success and solve for the mean failure rate that must be achieved in order to meet this standard.

Example

A respirator must be designed to achieve a 90% probability of no failures in 1000 hours of use. What test rate (λ) must be experienced to meet this requirement.

Solution

$$P(r) = \frac{(\lambda t)^r e^{-\lambda t}}{r!}$$

where r = 0
t = 1000
$P(r)$ = .9 (our desired probability for $P(0)$)
λ = unknown variable

Thus,

$$.9 = \frac{(\lambda * 1000)^0 e^{-\lambda * 1000}}{0!} = e^{-\lambda * 1000}$$

To eliminate the superscript ($-\lambda * 1000$), we can take the natural log of both sides of the equation. We get

$$\ln(.9) = -\lambda * 1000 \ln e$$

Now use a scientific calculator to solve for λ. (By definition, $\ln e = 1$):

$$-0.1053605 = -\lambda * 1000 * 1$$

$$\frac{-0.1053605}{1000} = \frac{-\lambda * 1000 * 1}{1000}$$

$$-1.053605\,\text{E-4} = -\lambda$$

$$1.053605\,\text{E-4} = \lambda$$

The answer means the respirator can have only slightly more than 1 failure in 10,000 hours of test. We would need an MTBF of 9491 hours ($1/\lambda = 1/.0001054$). But remember that the Poisson distribution formula cannot distinguish between respirators. Therefore, the design is only for a 90% probability of 1000 hours *total time without a failure*, not for one respirator working 1000 hours 90% of the time.

Example

Now let's examine another question. Suppose we are already manufacturing a respirator with an MTBF of 800 hours and we want to tell customers how long they can expect the respirator to operate with 0 failures 85% of the time.

Solution   We use the Poisson distribution:

$$P(r) = \frac{(t/m)^r e^{-t/m}}{r!}$$

where r = 0
   m = 800
   $P(0)$ = .85 (our given probability)
   t = unknown time respirator should operate

Substituting these values in the equation gives

$$.85 = \frac{(t/800)^0 e^{-t/800}}{0!} = e^{-t/800}$$

Taking natural logs of both sides of the equation yields

$$\ln(.85) = \frac{-t}{800}\ \ln e$$

Using a scientific calculator, we find

$$-0.16251893 = \frac{-t}{800} * 1$$

$$-0.16251893 * 800 = \frac{-t}{800} * 800$$

$$-130.0151 = -t$$

$$t = 130.0151$$

We could therefore tell customers that 85% of the time our respirators operate for over 130 hours without a failure. Thus, if the customer purchased one respirator, he or she could expect it to work for 130 hours without a failure 85% of the time, but if two respirators were purchased, the 130 hours must be shared between them, so the customer should not expect the respirators to last over 65 hours. It really poses an awkward situation for analysis. Therefore, be careful how you use the Poisson distribution.

As long as you know any three of the four variables in the Poisson formula (λ, t, r, $P(r)$), you can solve for the remaining one.

You should now be ready to attempt the sample problems.

SAMPLE PROBLEMS

6.1. An airline company experienced 15 engine failures in 12,000 flight hours. What is the probability of no failures in the next 1000 flight hours?

a. 3.581 E-1   (230)
b. 3.490 E-1   (277)
c. .2865   (242)
d. .7135   (221)

6.2. For Problem 6.1, what is the probability of exactly 2 engine failures in 1500 flight hours?

a. .2238   (174)
b. .2875   (217)
c. .2044   (183)
d. .2696   (186)

6.3. For Problem 6.1, what is the probability of more than 1 engine failure in 700 flight hours?

a. .3648   (179)
b. .6352   (239)
c. .2184   (226)
d. .7816   (262)

6.4. A fire suppression system is tested and has an MTBF of 900 hours. What is the probability of any failures in the next 300 hours of operation for 2 systems (300 hours each)?

    a. .4866   (274)
    b. .4883   (205)
    c. .4868   (192)
    d. .7769   (177)

6.5. For the data in Problem 6.4, what is the probability of less than 2 failures?

    a. 8.557 E-1   (172)
    b. 1.443 E-1   (249)
    c. 3.423 E-1   (235)
    d. 5.134 E-1   (265)

6.6. Twelve systems were each tested for 150 hours. During this test, there were 4 failures. What is the probability of three failures in the next 140 hours?

    a. 3.5824 E-2   (198)
    b. 2.1361 E-2   (201)
    c. 3.6769 E-3   (199)
    d. 2.2241 E-1   (184)

6.7. If the desired probability of any failures in a system in 100 hours is .25, what must be the true MTBF of the system?

    a. 71   (258)
    b. 72   (253)
    c. 73   (247)
    d. 74   (240)

7

Normal Distribution

The normal distribution is probably the most abused distribution in statistics. At times, unskilled, would-be statisticians assume normalcy without ensuring that the rules and requirements for normalcy are met. However, for practical purposes, from a safety standpoint, relative accuracy with the normal distribution is achieved if large enough samples for data are used. Usually if more than 30 data points are in the sample, the normal distribution will be reasonably accurate.

The normal distribution probability formula has several advantages. By definition, the mean, median, and mode of a normal distribution are located at the same point. That is, the point representing the arithmetic average of the data (mean), the point at which exactly half of the values fall to the left and half fall to the right (median), and the value that occurs the most often (mode) are all the same data point. See Figure 7-1. This fact allows us to determine several things based on the information available.

The formula generally used to solve problems involving a normal distribution is

$$z = \frac{x - \bar{x}}{s}$$

where z = value representing an area under the curve
x = point of interest
$\bar{x}$ = arithmetic mean of a sample
s = standard deviation of a sample

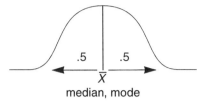

Figure 7-1

The mathematical symbols for a sample population are used because you will rarely have all of the data from a population. If population data is available, the formula is the same except that μ and σ are used in place of $\bar{x}$ and s. The only real difference is the computation of standard deviation.

The z represents an area under the normal curve, to be determined by using a table of normal distributions. There are many versions of these tables, but the correct use of the one available to you will result in the same answer. However, one problem is that, in Table B.1 at the back of this guide, all values of z are considered positive (the sign is ignored in the table). Before we clarify this issue, we will discuss how to use Table B.1.

To use Table B.1 you must remember that z merely represents a value that allows you to determine the area under the normal curve. Due to the properties of the normal distribution (mean = median = mode), the area under the curve is equivalent to a probability of occurrence. **Caution:** The area on various z tables is not always the area between the same points. In Table B.1 the area represented is the area between $\bar{x}$ (the mean) and x (the point of interest). In this guide, this area is denoted z_{rep} (see Figure 7-2).

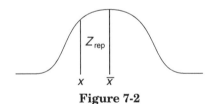

Figure 7-2

To use Table B.1, first determine the value of z (sometimes it is given to you, and sometimes you solve for it). Then look down the left column, which is the first two digits of z (the integer and the tenths). The values across the top represent the second decimal point of the z value (hundredths). Some examples may help clarify this.

Example

For $z = .9$, find z_{rep}.

Solution   Look down the left column of Table B.1 until you find .9. Since there is no value in the hundredths, look at the top row of the table and find 0.00. Follow that column down until it meets the row for .9. The value at this intersection is the value for the area between the mean and the point of interest (z_{rep}). Thus z_{rep} = 0.31594.

Example

If z = 2.9, find z_{rep}.

Solution   Look down the left column of Table B.1 until you find 2.9. Since there is no value in the hundredths, look at the top row of the table and find 0.00. Follow that column down until it meets the row for 2.9. The value at this intersection is the value for the area between the mean and the point of interest (z_{rep}). Thus, z_{rep} = 0.49813.

Example

If z = 1.85, find z_{rep}.

Solution   Look down the left column of Table B.1 until you find 1.8. Then look at the top row of the table to find 0.05. Follow that column down until it meets the row for 1.8. Thus, z_{rep} = 0.46784.

Example

If z = 1.643, find z_{rep}.

Solution   Look down the left column of Table B.1 until you find 1.6. Then look at the top row of the table to find 0.04. Follow that column down until it meets the row for 1.6. The value is 0.44950. Now we must account for the thousandths that were given in the value of z. The method to do this is called interpolation. To interpolate, we now find the value for 1.65, which is 0.45053. Now 1.643 is .3 of the way between 1.64 and 1.65 (likewise, 1.2356 is .56 of the distance between 1.23 and 1.24). Therefore we must calculate the area represented by the .3 difference between 1.64 and 1.65. To do this, subtract the z_{rep} for 1.64 from the z_{rep} for 1.65: 0.45053 – 0.44950 = 0.00103. Multiply it by .3 because this determines .3 of the distance between 1.64 and 1.65: .3 $*$ .00103 = 0.000309. This value must be added to the z_{rep} of 1.64. The final result is 0.449809. This represents the value for the area between the mean and the point of interest (z_{rep}) for a z of 1.643.

Example

An example without so many words may be useful. Determine the area between the mean and the point of interest (z_{rep}) if z = 1.867.

Solution   For z = 1.867, z_{rep} (1.86) = 0.46855 and z_{rep} (1.87) = 0.46925. The distance needed between 1.86 and 1.87 = .7. Thus

$$z_{rep}\,(1.87) - z_{rep}\,(1.86) = .46925 - .46855 = .0007$$
$$.7 * .0007 \qquad\qquad = .00049$$
$$z_{rep}\,(1.867) \qquad\qquad = 0.46855 + .00049 = .46904$$

Interpolation can be used for any value of z. If the value of z is 1.8679, the only difference from the preceding example is that the .0007 would be multiplied by .79 and that result would be added to 0.46855 ($z_{rep}\,(1.8679) = 0.469103$).

Now that we can read the table, how do we use it? Three types of questions can be asked: (1) What is the probability of a value being less than the point of interest? (2) What is the probability of a value being greater than the point of interest? (3) What is the probability of a value lying between two points of interest? Because of the unique properties of the normal distribution, all three questions are relatively easy to answer.

Remembering that exactly half of the curve lies to the left of the mean and half lies to the right allows us to use simple arithmetic to answer these questions. Question (1) has two possible solutions. If the point of interest is less than the mean, subtract z_{rep} from .5. The reason is that half the area lies to the left of the mean, and z_{rep} is the area between the mean and the point of interest, so the desired probability $(<x)$ lies to the left of x and is found by removing the area between x and the mean from the total area to the left. This is a lot of words, and a picture probably makes it clearer. Figure 7-2a represents this picture.

The same logic applies to the various combinations, and they are presented pictorially in Figures 7-2b–d.

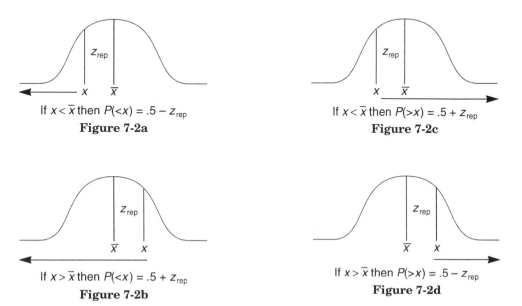

If $x < \bar{x}$ then $P(<x) = .5 - z_{rep}$
Figure 7-2a

If $x < \bar{x}$ then $P(>x) = .5 + z_{rep}$
Figure 7-2c

If $x > \bar{x}$ then $P(<x) = .5 + z_{rep}$
Figure 7-2b

If $x > \bar{x}$ then $P(>x) = .5 - z_{rep}$
Figure 7-2d

The easiest solution is for the probability of occurrence between two points of interest. See Figures 7-2e and 7-2f.

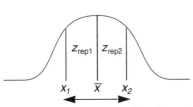

z_{rep1} z_{rep2}

x_1 $\overline{x}$ x_2

If x_1 and x_2 are on opposite sides of $\overline{x}$, then P(between x_1 and x_2) = z_{rep1} + z_{rep2}

Figure 7-2e

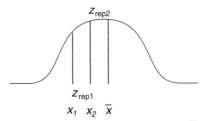

z_{rep2}

z_{rep1}

x_1 x_2 $\overline{x}$

If x_1 and x_2 are on same side of $\overline{x}$, then P(between x_1 and x_2) = z_{rep1} $-z_{rep2}$

Figure 7-2f

The following examples use these guidelines. In these problems assume that the normal distribution has been determined to be appropriate. It is also suggested that, rather than trying to memorize the above combinations of solutions, you draw a graph for each solution.

Example

If a production run of restraint harnesses was sampled and the mean failure was 2000 psi with a standard deviation of 40, what is the probability of a harness failing at less than 1950 psi?

Solution

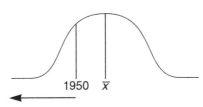

1950 $\overline{x}$

Figure 7-3

$$z = \frac{1950 - 2000}{40} = \frac{50}{40} = 1.25$$

where z = value representing an area under a normal curve
       x = point of interest = 1950 psi
       $\overline{x}$ = arithmetic mean of a sample = 2000 psi
       s = standard deviation of a sample = 40 psi

$$z_{rep} = 0.39436$$
$$P(<x) = .5 - .39436 = .1056400$$

This means that less than 10.564% of the harnesses will fail before 1950 psi.

Example

Assuming the same mean failure of 2000 psi with a standard deviation of 40, what is the probability that a harness will sustain a load of 2050 psi?

Solution

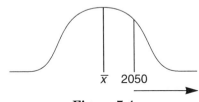

Figure 7-4

$$z = \frac{2050 - 2000}{40} = \frac{50}{40} = 1.25$$

where z = value representing an area under a normal curve
  x = point of interest = 2050 psi
  $\bar{x}$ = arithmetic mean of a sample = 2000 psi
  s = standard deviation of a sample = 40 psi

$$z_{rep} = 0.39436$$
$$P(>x) = .5 - .39436 = .1056400$$

   This problem is really asking the probability of a harness sustaining loads greater than 2050 psi without failing. This is more easily seen from the figure. The probability of withstanding loads greater than 2050 psi is the same as the probability of failing prior to 1950 psi, due to the symmetry of the normal distribution. Another way of looking at this result is that there is an 89.436% probability that the harness will fail before 2050 psi.

Example

Using the same data, what is the probability of failure between 1950 psi and 2050 psi?

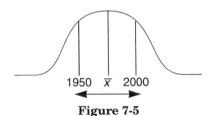

Figure 7-5

$$z_{rep}(1950) = 0.39436 \qquad z_{rep}(2050) = 0.39436$$
$$P(x_1 < \bar{x} < x_2) = 0.39436 + .39436 = .78872$$

It was not necessary to re-solve for the z_{reps}, so we only needed to sketch the picture to determine that we needed to add the z_{reps}.

Example

If a pressure relief valve on a boiler fails to function at 410 psi, the boiler could explode. Tests have demonstrated that the valves have a mean function rate of 390 psi with a standard deviation of 6 psi. However, because of process control, the relief valve must not function before 380 psi. What is the probability that a valve will not operate before 380 psi?

Solution

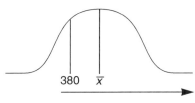

$$380 \quad \bar{x}$$

Figure 7-6

$$z = \frac{380 - 390}{6} = \frac{10}{6} = 1.667$$

where z = value representing an area under a normal curve
$\quad x$ = point of interest = 380 psi
$\quad \bar{x}$ = arithmetic mean of a sample = 390 psi
$\quad s$ = standard deviation of a sample = 6 psi

$$z_{rep} = 0.45224$$
$$P(<x) = .5 + .45224 = .95224$$

Interpolation was required to determine z_{rep} as follows: $z_{rep}(1.67) - z_{rep}(1.66) = .45254 - .45154 = .001$. Then $.001 * .7 = .0007$, which was added to $z_{rep}(1.66)$: $.0007 + .45154 = .45224$.

Example

Using the same data, what is the probability of the valve operating prior to an explosion due to overpressure?

Solution

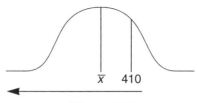

Figure 7-7

$$z = \frac{410 - 390}{6} = \frac{20}{6} = 3.333$$

where z = value representing an area under a normal curve
x = point of interest = 410 psi
$\bar{x}$ = arithmetic mean of a sample = 390 psi
s = standard deviation of a sample = 6 psi

$$z_{rep} = 0.499566$$
$$P(<x) = .5 + .499566 = .999566$$

Interpolation was required to determine z_{rep} as follows: $z_{rep}(3.34) - z_{rep}(3.33) =$ .49958 − .49956 = .00002. Then .00002 * .3 = .000006, which was added to $z_{rep}(1.66)$: .000006 + .49956 = .499566.

So far, we have known z or calculated it based on test data. The normal distribution can also be used in design and management. Let's use three examples to demonstrate.

Example

Assuming a normal distribution, suppose that an aircraft's engine has a mean time between failure (MTBF) of 1000 hours with a standard deviation of 20 hours. Find the number of hours that an engine can be expected to work with an 85% probability that it will operate that number of hours without failing?

Solution At first this problem may seem impossible, but a sketch and a little thought provide a solution. First, we cannot solve the equation for z because x is unknown. However, by sketching the curve we realize that we want the area at the far left of the curve to equal no more than .15. This means that the area between the mean and x must be .35. Now you can go to Table B.1 and, using it backwards, determine z. In the table, find the closest z_{rep} to .35 without going under .35. (It is not necessary to interpolate, although you could do so.) This

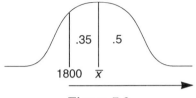

Figure 7-8

value is .34850, which is the area for a z of 1.03. Now that we know z, the mean, and the standard deviation, the only unknown is x, which is what we want to find. Hence,

$$P(>x) = .5 + .35 = .85$$

$$z_{rep} = 0.35$$

$$z = 1.03$$

$$-1.03 = \frac{x - 1000}{20}$$

where z = value representing z_{rep} = 1.03
x = unknown point of interest
$\bar{x}$ = arithmetic mean of sample = 1000 psi
s = standard deviation of sample = 20 psi

$$-1.03 * 20 = \frac{x - 1000}{20} * 20$$

$$-20.6 = x - 1000$$

$$-20.6 + 1000 = x - 1000 + 1000$$

$$979.4 = x$$

This means we can expect 85% of the engines to last more than 979.4 hours.

We need to clarify when z is positive and when it is negative. When $x < \bar{x}$, z is negative. If $x > \bar{x}$, then z is positive. Therefore, determining on which side of $\bar{x}$ that x lies will determine the sign of z.

Another use of the normal distribution is to determine a tolerance level for production. For example, as long as a desired mean has been set and the probability of maintaining a specific x has been established, the equation can be used to determine the standard deviation that must be achieved to reach that goal.

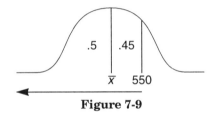

Figure 7-9

Example

Suppose that a pressure release valve has a mean function point of 500 psi. If it is desired to have a production line maintain that function point with a probability of .95 that the valves will operate at less than 550 psi, what is the value that the standard deviation of the valves must not exceed?

Solution

$$P(<x) = .5 + .45 = .95$$

$$z_{\text{rep}} = 0.45$$

$$z = 1.64$$

where z = value representing area under normal curve = 1.64
x = point of interest = 550
$\bar{x}$ = arithmetic mean of sample = 500 psi
s = unknown standard deviation of sample

$$1.64 = \frac{550 - 500}{s}$$

$$1.64 * s = \frac{550 - 500}{s} * s$$

$$1.64s = 550 - 500$$

$$1.64s = 50$$

$$s = \frac{50}{1.64} = 30.49$$

Again we determine z by working backwards in Table B.1. This time, however, since we are finding a probability that is less than an x that is greater than $\bar{x}$, we want to use the number closest to .45 without going over it. This value is .44950, which is the area for a z of 1.64. In this case z is positive because $x > \bar{x}$. Then we solve for s.

As long as all other variables are specified, we can solve the equation to determine the standard deviation for the mean.

Example

Suppose we desire to build an early warning system that will function 93% of the time for at least 100 hours with a standard deviation of 4 hours. What is the required MTBF?

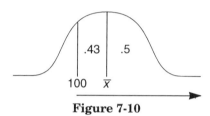

Figure 7-10

$$P(>x) = .5 + .43 = .97$$

$$z_{rep} = 0.43$$

$$z = 1.48$$

where z = value representing area under normal curve = -1.48

x = point of interest = 100

$\bar{x}$ = unknown arithmetic mean of sample

s = standard deviation of sample = 4 hours

$$-1.48 = \frac{100 - \bar{x}}{4}$$

$$-1.48 * 4 = \frac{100 - \bar{x}}{4} * 4$$

$$-5.92 = 100 - \bar{x}$$

$$-5.92 - 100 = 100 - \bar{x} - 100$$

$$-105.92 = -\bar{x}$$

$$\bar{x} = 105.92$$

The major difficulty in this problem is determining z's sign. From the figure we realize that since .93 of the area under the curve must be to the right of x, then x must be to the left of z. Thus, z is negative and the value found in the table is .43000. Again we want the closest value without going under .43000.

This means that our production standard must produce alarms that have an MTBF of 105.92 hours to ensure that 93% of them would last 100 hours.

The normal distribution allows us to solve a variety of problems. But you need to learn when the normal distribution applies and when you should not assume it.

SAMPLE PROBLEMS

7.1. A smoke alarm has been determined to have a mean life of 1400 days with a standard deviation of 50 days. Assuming that the life is normally distributed, what is the probability of an alarm failing before 1380 days?

      a. .68918   (231)
      b. .655410   (251)
      c. .34459   (174)
      d. .15541   (188)

7.2. For the data of Problem 7.1, what is the probability of an alarm not failing before 1462 days?

      a. .10748   (215)
      b. .89252   (218)
      c. .39252   (225)
      d. .24810   (261)

7.3. For the data of Problem 7.1, what is the probability of an alarm lasting between 1300 and 1500 days?

      a. .47724   (235)
      b. .97724   (243)
      c. .02276   (248)
      d. .95448   (256)

7.4. For the data of Problem 7.1, let the standard deviation be 52. What is the probability of an alarm failing before 1500 days?

      a. .972749   (278)
      b. .027251   (271)
      c. .472749   (267)
      d. .743812   (264)

7.5. A harness must have a 95% chance of withstanding 1000 psi without failing. If manufacturing techniques provide a constant standard deviation of 30 psi, what is the mean psi of failure that must be achieved?

      a. 1049.2   (213)
      b. 1049.5   (206)
      c. 990.1   (202)
      d. 990.4   (189)

7.6. To ensure that a flight control system lasts 2200 hours 40% of the time, with a mean time to failure of 2100, what is the level of deviation that can be allowed?

a. 360.4   (185)
b. 384.6   (177)
c. 400.0   (171)
d. 420.2   (245)

8

Chi-Square Confidence Intervals

Suppose you have five years of experience to show that your probability of an accident is .005. Next year you expect to make 1000 trips. Therefore, you could logically expect to have five accidents (.005 ∗ 1000). However, how likely is it that you will really experience exactly five accidents? True, you can compute the probability of five accidents in 1000 trips, but let's look at the problem from another perspective. Rather than going to management and telling them that there is a .1759076 probability (based on the binomial formula for $P(5/1000,.005)$, might it not be just as good or even better to go to management and state that you are 95% confident that you will experience no more than a specific amount of accidents? This is the concept behind confidence intervals. Let's use a nonsafety example to help understand this concept.

If you weighed 10 employees today and discovered that their average weight was 160 pounds, what would you expect the average weight to be if you weighed them tomorrow? The obvious answer is 160.0 pounds, but do you really believe that it would be exactly 160.0 pounds? Suppose 1 of the 10 gorged to such a degree that 5 pounds were added to his or her weight. Even if each of the other nine weighed exactly the same as the day before, the average weight would be increased to 160.5 pounds. On the next day what would you expect the weight to be? By now you probably have the idea. In reality it is extremely unlikely that you can measure anything at different times and achieve identical results. So what is the true mean weight of the employees? There is no such thing as the "true" mean.

To deal with this problem, statisticians developed the concept of confidence intervals. There are many different ways to determine confidence intervals, depending on what type of distribution the data represent—normal, t, chi-square (χ^2), binomial, and so on. The principle is the same for all of them: we want to determine a range for the mean to ensure that we can be confident to a specific degree that the mean falls within that range. For example, if we want a 90% confidence interval, then the sampled mean result will fall within that range 90% of the time. In other words, if we conduct an experiment and determine the mean weight of 10 people 100 times, 90 of the mean weights will be within the confidence interval calculated.

If the mean is exponentially distributed, it has been demonstrated that the chi-square distribution can approximate the mean time between failure. For the purposes of this guide, we will examine only the χ^2 confidence interval. Certain concepts should be understood before we proceed.

First, the χ^2 distribution is partially determined by the number of independent events. These events help determine the degrees of freedom (v). Without discussing the theory behind it, v drives the shape of the distribution. Also of importance, when $v > 30$, the χ^2 distribution resembles the normal distribution, which influences the χ^2 formula.

The second concept concerns the range. The range for mean time between failure (MTBF) can be developed in one of two ways. The first is called a two-tailed test. A two-tailed test allows the mean to fall between two numbers. An example might be an alarm system. In this instance, from a safety perspective, we want to ensure that the alarm does not have a failure rate that is too high. A production engineer, however, does not necessarily want to overengineer the alarm to drive the cost tremendously high just to obtain an unnecessarily low failure rate. Therefore, we are interested in determining a confidence interval between two extremes. This means that we must account for error at both ends of the distribution. Error is denoted by α. As in all probabilities, the sum of the probabilities of everything that can happen must equal 1. In our example we can be correct or in error. Being correct will be called our confidence interval (CI). Therefore,

$$\text{CI} + \alpha = 1 \qquad \text{or} \qquad 1 - \text{CI} = \alpha$$

Thus, if we want a 95% confidence, α is .05. However, for a two-tailed test, we are concerned about erring at both ends of the distribution, so α must be divided by 2.

In the example, safety was concerned with only one end of the spectrum. If we ignore production's concerns, we are concerned about erring in only one direction, and α need not be divided. This means that this test is conservative because the error is all on one end. Generally speaking, safety is concerned only if there are more failures than anticipated. Therefore, safety's concern is normally with the upper bound.

Because of the various combinations of concern, basically three formulae for χ^2 confidence intervals exist. They are

$$P\left(\frac{\chi^2_{1-\alpha/2;2F}}{2T} < \lambda < \frac{\chi^2_{\alpha/2;2F+2}}{2T} \right) = 1 - \alpha \quad \text{two-tailed test}$$

$$P\left(\lambda < \frac{\chi^2_{\alpha;2F+2}}{2T} \right) = 1 - \alpha \quad \text{upper one-tailed test}$$

where λ = failure rate
      α = error
      F = historical failures, faults, accidents, etc.
      T = historical trials, time, trips, exposures, etc.
      $2F$ or $2F + 2$ = v = degrees of freedom

The formula to use depends on the information sought. If you are concerned about failures at both ends of the spectrum, too early or too late, then the two-tailed test must be used. If you are interested in limiting the error to the upper end of the spectrum, the one-tailed test is needed. Based on the nature of the problem presented in the workplace, you should be able to determine which formula to use. A χ^2 table (Table A.2) is needed to determine the confidence interval.

To use Table B.2 simply determine the error rate and degrees of freedom. The error rate is given at the top of the table. The error rate is determined by the value in the subscript before the semicolon. For example, for $\chi^2_{.05;8}$, .05 is the error rate. The degrees of freedom (v) are displayed down the left column of the table. For $\chi^2_{.05;8}$, the degrees of freedom are 8. Where these two values intersect is the value of $\chi^2_{.05;8}$, 15.507.

Example

A safety officer is interested in the true failure rate for a safety device sensing unit. The unit has experienced 10 failures in 1000 hours of test. What is the 95% confidence interval?

Solution   A safety officer wants to ensure that the failure rate is not worse than anticipated. Therefore, the formula for a one-tailed upper bound should probably be used:

$$P\left(\lambda < \frac{\chi^2_{\alpha;2F+2}}{2T}\right) = 1 - \alpha$$

where α = .05, F = 10, and T = 1000. Substituting the values for the formula yields

$$\lambda < \frac{\chi^2_{.05;2(10)+2}}{2(1000)}$$

Finding the value for the $\chi^2_{.05;22}$ yields

$$\frac{33.924}{2000} < .016962$$

Hence, although the tested failure rate is .01 (10/1000), based on the confidence interval, the safety officer can be 95% sure that the real failure rate is less than .016962 or that in 100 trials 95 of them would be expected to result in a λ less than .016962.

Example

Now assume that production engineering is concerned that if the MTBF is too low, they will spend too much on engineering. The unit has still experienced 10 failures in 1000 hours of test. What is the 50%, two-tailed confidence interval?

Solution The question states that production wants to know the 50% confidence interval including both tails, so the formula for a two-tailed test is needed:

$$P\left(\frac{\chi^2_{1-\alpha/2;2F}}{2T} < \lambda < \frac{\chi^2_{\alpha/2;2F+2}}{2T}\right) = 1 - \alpha$$

where $\alpha = .5$, $F = 10$, $T = 1000$, λ = unknown. Substituting the values for the formula yields

$$\frac{\chi^2_{1-.5/2;2(10)}}{2(1000)} < \lambda < \frac{\chi^2_{.5/2;2(10)+2}}{2(1000)}$$

Finding the values for the $\chi^2_{.75;20}$ and $\chi^2_{.25;22}$ yields

$$\frac{15.452}{2000} < \lambda < \frac{26.039}{2000}$$

Hence, although the tested failure rate equals .01 (10/1000), based on the confidence interval, production can be 50% sure that the real mean is greater than .007726 and less than .0130195. Since $\lambda = 1/m$, this value equates to an MTBF between 129.43 and 76.81 hours.

What do we do if there are more than 30 degrees of freedom? Table B.2 stops at 30. Above 30 degrees of freedom, χ^2 distributions resemble the normal distribution. Therefore, a footnote to the table explains the procedure when there are more than 30 degrees of freedom. The formula for this χ^2 is

$$\chi^2 = .5 * (z_\alpha + (2v-1)^{.5})^2$$

where z_α = value in table under appropriate error rate
v = degrees of freedom

After χ^2 is determined, problems with more than 30 degrees of freedom are computed just as for any χ^2 confidence interval problem.

Example

A safety officer is interested in the true failure rate for a smoke alarm unit. The unit has experienced 15 failures in 4000 hours of test. What is the 90% confidence interval?

Solution   The safety officer wants to ensure that the failure rate is not worse than anticipated. Therefore, the formula for a one-tailed upper bound is used:

$$P\left(\lambda < \frac{\chi^2_{\alpha;2F+2}}{2T}\right) = 1-\alpha$$

where $\alpha = .1$, $F = 15$, $T = 4000$, and λ = unknown. Substituting the values for the formula yields

$$\lambda < \frac{\chi^2_{.1;2(15)+2}}{2(4000)}$$

Since the value for $\chi^2_{.1;32}$ cannot be found in the table it must be calculated from the formula

$$\chi^2 = .5 * (z_\alpha + (2v-1)^{.5})^2$$

Substituting the values for z_α and v yields

$$
\begin{aligned}
\chi^2 &= .5 * (1.282 + (2 * 32 - 1)^{.5})^2 \\
&= .5 * (1.282 + (64 - 1)^{.5})^2 \\
&= .5 * (1.282 + (63)^{.5})^2 \\
&= .5 * (1.282 + 7.9372539)^2 \\
&= .5 * (9.219253933)^2 \\
&= .5 * 84.99464309 \\
&= 42.49732
\end{aligned}
$$

$$
\begin{aligned}
\lambda \quad &< \frac{\chi^2_{.1;2(15)+2}}{2(4000)} \\
&< \frac{42.49732}{8000} \\
&< .005312165
\end{aligned}
$$

Hence, although the tested failure rate equals .00375 (15/4000), based on the confidence interval, the officer can be 95% sure that the real failure rate is less than .005312165.

Because of the numerous steps involved in computing χ^2 when the degrees of freedom are greater than 30, it is possible to make a mistake in computation. Common mistakes in computing χ^2 include using F instead of the actual number of degrees of freedom, failing to square $(z_\alpha + (2v-1)^{.5})$, and failing to divide the product by 2.

You can check to see if your computed χ^2 is reasonable. Let's use the above computation as an example. First, find the highest value for χ^2 for the error rate of interest, 40.256 ($\chi^2_{.1;30}$) in this example. Now determine how many degrees of freedom more than 30 you used. In this example, $32 - 30 = 2$. Next find the χ^2 which is that much less than the highest χ^2 found. Since the highest χ^2 was 40.256, the χ^2 that is 2 degrees of freedom less than that is 37.916 ($\chi^2_{.1;28}$). The difference between the two is about 2.3 ($40.2 - 37.9$). Therefore, your computed χ^2 should be about that much larger than the 40.256. Isn't 42.49 about 2.3 greater than 40.255? So χ^2 is probably computed accurately.

Confidence intervals can also be used to determine design specifications. For example, you might determine the number of failures allowed within specific testing exposures in order to have a specific confidence level.

Example

Using 800 hours of testing time, you want to ensure a λ of no less than .02 with 97.5% confidence. What is the maximum number of failures you can experience within the test?

Solution   The information in the problem leads you to believe you would probably use the formula for a one-tailed upper bound:

$$P\left(\lambda < \frac{\chi^2_{\alpha;2F+2}}{2T}\right) = 1-\alpha$$

where $\alpha = .025$, F = unknown, $T = 800$, and $\lambda = .02$. Substituting the known values for the formula yields

$$.02 < \frac{\chi^2_{.025;2F+2}}{2(800)}$$

$$.02 < \frac{\chi^2_{.025;2F+2}}{1600}$$

$$.02 * 1600 < \frac{\chi^2_{.025;2F+2}}{1600} * 1600$$

$$32.0 < \chi^2_{.025;2F+2}$$

To solve for F, find the value for .025 in the table that is closest to 32 without going over. A value greater than 32 would cause the failure rate to be greater than .02. The closest value is 31.526. This occurs at $\chi^2_{.025;18}$. Therefore, $2F + 2 =$ 18, or $F = 8$. Thus, to ensure a failure rate of less than .02 in 800 hours of test, with a confidence of .975, you could not experience more than 8 failures.

Any variable could be determined in the same manner. For example, holding F, m, and CI constant, you could solve for the required T.

Another use for determining the CI leads back to our original problem in this chapter. After a confidence interval is found, it is possible to use the Poisson formula to determine the confidence you have in the probability you determine.

Example

Assume that you have experienced 6 failures of a fire-sensing mechanism in 850 tests. What is the probability that the sensing mechanism will complete, with a 95% confidence, 100 missions without a failure?

Solution

$$P\left(\lambda < \frac{\chi^2_{\alpha;2F+2}}{2T}\right) = 1 - \alpha$$

where $\alpha = .05$, $F = 6$, $T = 850$, and λ = unknown. Substituting the values for the formula yields

$$\lambda < \frac{\chi^2_{.05;2(6)+2}}{2(850)}$$

$$< \frac{\chi^2_{.05;2(6)+2}}{1700} < \frac{\chi^2_{.05;14}}{1700}$$

$$< \frac{23.685}{1700} < .013932353$$

Substitute λ in the Poisson formula to obtain the answer:

$$P(0) = e^{-.013932353 \,*\, 100} = .248270775$$

Thus, we can state that we are 95% confident that there is a 24.83% probability that a sensing mechanism will have no failures in 100 uses.

SAMPLE PROBLEMS

8.1. If you want a 90% confidence interval, what error are you willing to accept?

a. .05   (230)
b. .1   (239)
c. .5   (246)
d. 1   (253)

8.2. In a two-tailed, upper-bound test, if the desired confidence is 90%, what error rate would you use in the table?

a. .5   (173)
b. .1   (179)
c. .05   (187)
d. .01   (190)

8.3. In a one-tailed, upper bound test, in the table, the error rate for a 90% confidence interval is

a. .01   (196)
b. .05   (199)
c. .1   (205)
d. .5   (210)

8.4. For a one-tailed, lower bound test, $F = 6$. What is the value of v?

a. 6   (215)
b. 12   (221)
c. 14   (227)
d. 3   (262)

8.5. For a two-tailed, upper bound test, $F = 10$. What is the value of v?

a. 10   (184)
b. 20   (275)
c. 22   (278)
d. 5   (273)

8.6. What is the value of $\chi^2_{.05;12}$?

a. 11.340   (270)
b. 21.026   (266)
c. 23.337   (264)
d. 36.415   (261)

8.7. If your process experienced 8 failures, what is the value of χ^2 for a 90% confidence interval of a two-tailed, upper bound?

    a. 23.542   (224)
    b. 25.989   (221)
    c. 26.296   (216)
    d. 28.869   (212)

8.8. If an 80% confidence interval is desired, what is the value of χ^2 for a one-tailed, upper bound if 11 failures are experienced?

    a. 29.553   (209)
    b. 27.301   (205)
    c. 33.196   (201)
    d. 30.813   (197)

8.9. What is the 95%, one-tailed, upper confidence bound on an item experiencing 12 failures in 1100 hours of test?

    a. .019091   (195)
    b. .0190559   (190)
    c. .0182332   (186)
    d. .017675   (181)

8.10. What is the 98%, two-tailed, upper confidence bound on an item experiencing 2 failures in 600 hours of test?

    a. .003333   (178)
    b. .011064   (172)
    c. .01401   (168)
    d. .02802   (255)

8.11. What is the 80%, one-tailed, upper confidence bound on an item experiencing 16 failures in 2000 hours of test?

    a. .00511625   (251)
    b. .0051340   (245)
    c. .0058855   (174)
    d. .0101844   (236)

8.12. If a desired 95%, one-tailed, upper-confidence-bounded rate is .015, how many additional tests must be made to a system experiencing 10 failures in 700 tests, to prove the desired rate if no additional failures occur?

    a. 221   (234)
    b. 431   (240)
    c. 429   (250)
    d. 220   (257)

9

Event Systems

A system consists of numerous subsystems arranged in various ways. The easiest system to analyze is one with its subsystems arranged in series. For a series system to be successful, flow from one subsystem to the next must occur linearly. In other words, each subsystem must function appropriately or the next subsystem will not receive its input. Thus, the probability of success relies on the Multiplication Law and appears as

$$P(\text{suc})_{\text{ser}} = P(s)_1 * P(s)_2 * P(s)_3 * \cdots * P(s)_n$$

Figure 9-1 represents an accident system. An accident system is a group of subsystems that work together in such a way that the outcome (success) of the system is an undesired event (accident). For example, Figure 9-1 represents the fire chain. Each subsystem is necessary, or there will be no fire. Likewise, if each subsystem functions as designed, there will be an undesired event—a fire. What is the probability of a fire in this system if the probability of success of each subsystem is the value noted below each block?

For the system to work, each part must work. Therefore, the probability of there being oxygen and fuel and ignition and reaction follows the Multiplication Law:

$$P(\text{Fire}) = P(\text{Oxy}) * P(\text{Fuel}) * P(\text{Ign}) * P(\text{React})$$
$$= .99 * .5 * .1 * .01 = .0004950$$

Note that the probability of system success is much smaller than the probability of success of the individual system components. Therefore, it is good to design

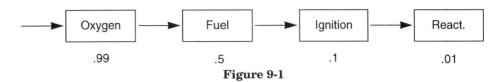

Figure 9-1

accident systems in series because the overall probability of success of the system will be much smaller than the probability of the components, and in accident systems we want the lowest possible probability of success.

The system can fail (no fire) in numerous ways. Any block failure causes system failure, but various combinations of blocks could also fail simultaneously. For example, perhaps oxygen and ignition are available but not fuel or reactivity. Rather than computing the probability for each of these combinations and using the Addition Law to compute the total probability of system failure, we can compute the probability of system failure (no fire) with the Complementary Law. Remember, this law states that the individual probabilities of all things that can happen must equal 1. Since the system can do only one of two things, succeed or fail, $P(\text{suc})_{\text{sys}}$ plus $P(\text{fail})_{\text{sys}}$ must equal 1:

$$
\begin{aligned}
P(\text{suc})_{\text{sys}} + P(\text{fail})_{\text{sys}} &= 1 \\
.0004950 + P(\text{fail}) &= 1 \\
P(\text{fail}) = 1 - .0004950 &= .9995050
\end{aligned}
$$

Another type of system is the reliability system. The successful outcome of a reliability system is a desired event. Figure 9-2 represents a reliability system for

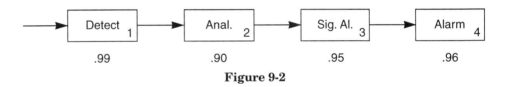

Figure 9-2

a simple smoke detector: First, the smoke must be detected by a device represented by block 1. Then an analyzer (block 2) analyzes the substance and determines that it is smoke. Block 3 is a device that sends a signal to the alarm (block 4), telling the alarm it needs to sound. Block 4 represents the actual functioning of the alarm. If each block works, the alarm sounds. If any block fails, the alarm remains silent. Therefore, the probability of detection, and analysis, and alarm signaled, and alarm sounded follows the Multiplication Law. Thus,

$$
\begin{aligned}
P(\text{suc})_{\text{alarm sounded}} &= P(s)_{\text{det}} * P(s)_{\text{anal}} * P(s)_{\text{sig}} * P(s)_{\text{alarm}} \\
&= .99 * .90 * .95 * .96 = .812592
\end{aligned}
$$

Note that the probability of system success is much smaller than the probability of success of the individual system components. Therefore, reliability systems should not be designed in series because we want the highest possible probability of success.

The probability of failure of the reliability system (no alarm) can also be computed from the Complementary Law:

$$P(\text{suc})_{\text{sys}} + P(\text{fail})_{\text{sys}} = 1$$
$$.812592 + P(\text{fail})_{\text{sys}} = 1$$
$$P(\text{fail})_{\text{sys}} = 1 - .812592 = .187408$$

How should reliability systems be designed? Let's try to improve the overall probability of success for the sample reliability system. The weakest link in the chain (system)—the analyzer—has a probability of successful analysis of only .9. If the individual analyzer's probability of success cannot be improved, is there a way to increase that block's probability of success? Suppose the problem is that the analyzer always makes the correct analysis ($P(\text{suc}) = 1$), but its success rate is only .9 because it works only 90% of the time. Now suppose we configured two analyzers in parallel so that if one didn't function the other might. See Figure 9-3.

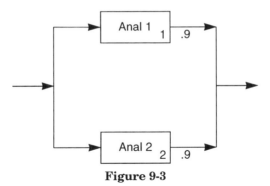

Figure 9-3

In a parallel system, the system's success does not depend on every subsystem's success because the system can work in numerous ways. In Figure 9-3, the number outside the block represents that block's probability of success. Since the analyzers are identical, the probability of success for each is the same, 90%. Note that even if analyzer 2 fails, but analyzer 1 works, the system still works. Thus, analysis can occur in several ways.

(a)   analyzer 1 works   analyzer 2 works
(b)   analyzer 1 works   analyzer 2 fails
(c)   analyzer 1 fails   analyzer 2 works

If the probability of analyzer 2 working is .9, then the probability of analyzer 2 failing is .1 (i.e., $1 - .9 = .1$). Thus, the probability of analyzer 1 working and analyzer 2 failing is $.9 * .1 = .09$. Table 9-1 lists these probabilities. Since the system works if (a), (b), or (c) occurs, the probability of the system working is the

TABLE 9-1. PROBABILITY OF SYSTEM FUNCTIONS (TWO PARALLEL)

(a)	analyzer 1 works	*	analyzer 2 works	=	.9	*	.9	=	.81
(b)	analyzer 1 works	*	analyzer 2 fails	=	.9	*	.1	=	.09
(c)	analyzer 1 fails	*	analyzer 2 works	=	.1	*	.9	=	.09

probability of (a) or (b) or (c) occurring. By the Addition Law, the probability of analysis taking place is then (a) + (b) + (c):

$$.81 + .09 + .09 = .99$$

Let's pause from this example for a minute to discuss parallel solutions in a bit more detail. The preceding solution is entirely correct, but there is an easier way to compute the probability of success using the Complementary Law. Since the system can either work (success) or fail, then $P(\text{suc}) + P(\text{fail}) = 1$. Analysis does not occur only if analyzer 1 and analyzer 2 both fail at the same time. This probability is

$$.1 * .1 = .01$$

Hence,

$$P(\text{suc}) + P(\text{fail}) \quad = 1$$
$$P(\text{suc}) + .01 \qquad = 1$$
$$P(\text{suc}) = 1 - .01 \ = .99$$

Notice that this is exactly the same answer as solving for all of the ways the system could work and adding them together. The importance of this easier solution can be seen by two more simple examples. Before attempting to improve the system in Figure 9-2, let's examine parallel systems further. Suppose three analyzers are in parallel (Figure 9-4). As long as one analyzer works, the system works. What is the probability of analysis? It could be determined as in Table 9-2. (For simplicity, only the block numbers are used. For example, 1w means that analyzer 1 works, and 2f means that analyzer 2 fails.) The probability of analysis is the sum of the individual probabilities:

$$(a) + (b) + (c) + (d) + (e) + (f) + (g) = .729 + .081 + .081 + .009 + .081 + .009 + .009 = .999$$

However, analysis can fail only if all three analyzers fail simultaneously. Hence, by the Complementary Law,

$$P(\text{fail})_{\text{sys}} = 1f * 2f * 3f = P(f)_1 * P(f)_2 * P(f)_3$$
$$= .1 * .1 * .1 = .001$$
$$P(\text{suc})_{\text{sys}} + P(\text{fail})_{\text{sys}} = 1$$
$$P(\text{suc})_{\text{sys}} = 1 - P(\text{fail})_{\text{sys}} = 1 - .001 = .999$$

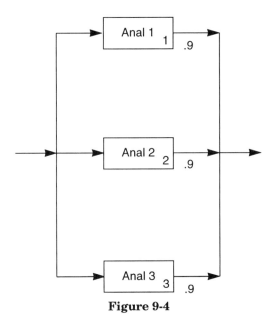

Figure 9-4

Tip: Instead of finding P(suc) and P(fail) for each subsystem whenever it's needed, it is easier to develop a table. For example, the following table can be developed for Figure 9-4:

Block	P(suc)	P(fail)
1	.9	.1
2	.9	.1
3	.9	.1

Consult your table each time you need a value.

Now suppose we have four analyzers in parallel (Figure 9-5). For more than three subsystems in parallel, the manual method becomes unwieldy and the only

TABLE 9-2. PROBABILITY OF SYSTEM FUNCTIONS (THREE PARALLEL)

(a)	1w	2w	3w	=	.9	*	.9	*	.9	=	.729
(b)	1w	2w	3f	=	.9	*	.9	*	.1	=	.081
(c)	1w	2f	3w	=	.9	*	.1	*	.9	=	.081
(d)	1w	2f	3f	=	.9	*	.1	*	.1	=	.009
(e)	1f	2w	3w	=	.1	*	.9	*	.9	=	.081
(f)	1f	2w	3f	=	.1	*	.9	*	.1	=	.009
(g)	1f	2f	3w	=	.1	*	.1	*	.9	=	.009

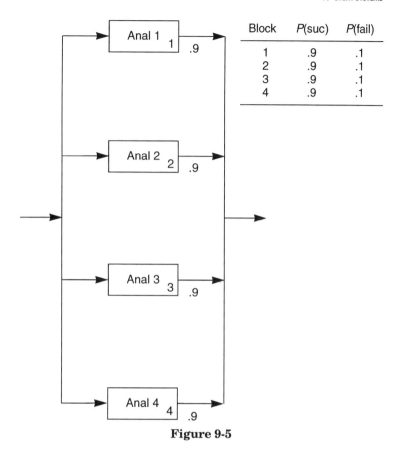

Block	P(suc)	P(fail)
1	.9	.1
2	.9	.1
3	.9	.1
4	.9	.1

Figure 9-5

easy way (without a computer) of obtaining the proper solution is by the Complementary Rule:

$$P(\text{suc}) + P(\text{fail}) = 1$$
$$P(\text{fail})_{\text{sys}} = 1\text{f} * 2\text{f} * 3\text{f} * 4\text{f} = P(f)_1 * P(f)_2 * P(f)_3 + P(f)_4$$
$$= .1 * .1 * .1 * .1 = .0001$$
$$P(\text{suc})_{\text{sys}} = 1 - P(\text{fail})_{\text{sys}} = 1 - .001 = .9999$$

Let's pause and examine the situation more carefully. Note that to solve for the probability of success $P(\text{suc})$ of any reliability system in parallel, you want to always use the complementary rule ($P(\text{suc}) + P(\text{fail}) = 1$). The general formula for a parallel system is then

$$P(\text{suc})_{\text{par}} = 1 - (P(f)_1 * P(f)_2 * P(f)_3 * \cdots * P(f)_n)$$

The probability of system success of a reliability system in parallel is much higher than the probability of success of each block in the system. Therefore, place components in a reliability system in parallel to increase the probability of system success.

Now let's consider an accident system in parallel, for example, the probability of injury in a ladder accident. Assume that there are three ways in which an injury might occur: (1) A person might fall from the ladder. (2) The ladder might slip and fall. (3) The wind might blow it over. Since any one of these three occurrences would result in an accident, they could be drawn as a parallel system. Figure 9-6 shows this system. The numbers outside each block represent the probability of that occurrence (success). This is sometimes difficult to comprehend, but just remember that the successful outcome of an accident system is an accident. Therefore, the concepts of success and failure are the opposite of what they normally seem to be.

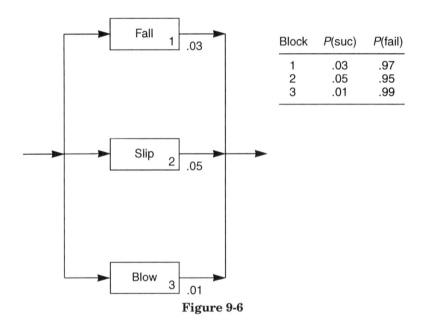

Block	P(suc)	P(fail)
1	.03	.97
2	.05	.95
3	.01	.99

Figure 9-6

Once more, the probability of success of a parallel system is

$$P(\text{suc})_{\text{par}} = 1 - (P(f)_1 * P(f)_2 * P(f)_3 * \cdots * P(f)_n)$$

or

$$= 1 - (.97 * .95 * .99) = 1 - .912285 = .087715$$

Note that the probability of system success (an accident) is substantially higher than the probability of success of any individual possibility of injury from the ladder. This is always true. Therefore, avoid designing accident systems in parallel.

Now returning to our original example for a reliability system in series (Figure 9-2), let's examine what the two analyzers have done to the reliability of the system. By redesigning block 2 to have two analyzers in parallel, the new design is given in Figure 9-7:

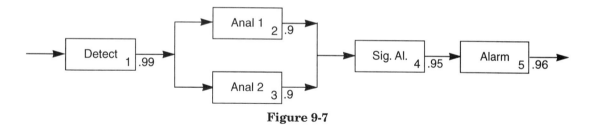

Figure 9-7

The probability of system success (an alarm) is now

$$P(s)_{sys} = P(s)_1 * (1 - (P(f)_2 * P(f)_3)) * P(s)_4 * P(s)_5$$
$$P(s)_{sys} = .99 * (1 - (.1 * .1)) * .95 * .96$$
$$P(s)_{sys} = .99 * .99 * .95 * .96 = .8938512$$

Two analyzers in parallel increased system reliability by more than 10% ((.8938512 − .812592)/.812592). Whenever possible, use parallel systems in reliability systems.

Solving for reliability by writing a formula directly from Figure 9-7 is not difficult, but what is the solution for the probability of success for a reliability system represented by Figure 9-8? Assume that it represents part of a rocket's flight system, where each block represents the reliability of the following:

Block	Function	Reliability (success)
1	Auxiliary power unit 1	.94
2	Auxiliary power unit 2	.94
3	Computer 1	.95
4	Computer 2	.96
5	Computer 3	.96
6	Engine 1	.97
7	Engine 2	.98
8	Flight controls	.999

The correct solution is

$$P(suc)_{sys} = (1 - P(f)_1 * P(f)_2) * (1 - (1 - P(s)_3 * P(s)_6))$$
$$* (1 - (1 - P(f)_4 * P(f)_5) * P(s)_7)) * P(s)_8$$

However, it is easy to make a mistake when writing the formula, especially by forgetting to subtract a number from 1. Therefore, why not attack the problem like eating an elephant (one bite at a time)?

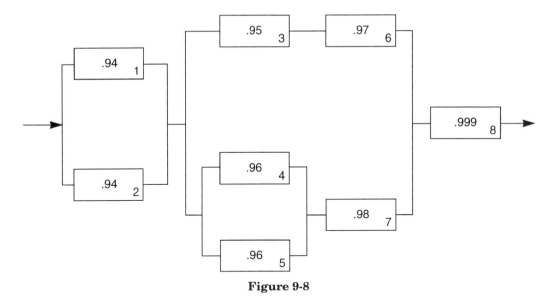

Figure 9-8

Let's separate the system into simple series and parallel systems. Figure 9-8 could be divided as shown in Figure 9-9.

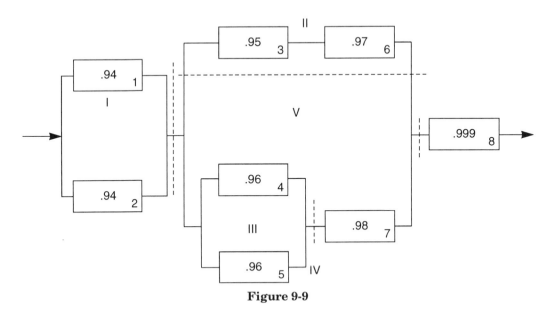

Figure 9-9

**TABLE 9-3. BLOCK PROBABILITIES
FOR FIGURE 9-9**

Block	P(s)	P(f)
1	.94	.06
2	.94	.06
3	.95	.05
4	.96	.04
5	.96	.04
6	.97	.03
7	.98	.02
8	.999	.001

Using the tip, we first make a table showing $P(s)$ and $P(f)$ of each block (Table 9-3).

Part I is two parallel blocks in series with the rest of the system. The probability of success of the APUs (blocks 1 and 2) is

$$P(s)_I = 1 - (P(f)_1 * P(f)_2)$$
$$= 1 - (.06 * .06) = 1 - .0036 = .996400$$

Although not necessary, redrawing the system helps keep things straight. See Figure 9-10.

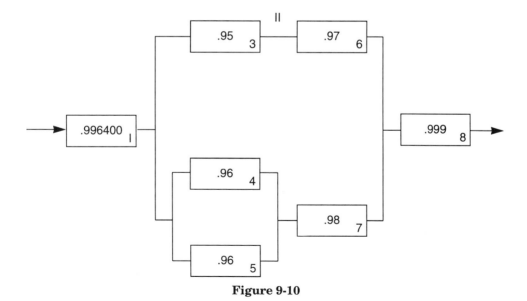

Figure 9-10

It is not necessary to make your drawing this detailed. It is for example purposes only.

Solving for the probability of success for Part II results in one block in parallel with IV. The probability of success of Comp3 and Eng2 is

$$P(s)_{II} = P(s)_3 * P(s)_6 = .95 * .97 = .921500$$

See Figure 9-11.

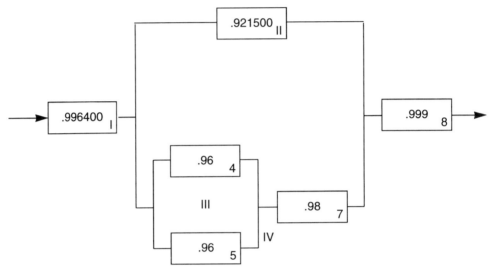

Figure 9-11

Solving for the probability of success for Part III results in one block in series with block 7. The probability of success of Comp1 and Comp2 is (Figure 9-12)

$$P(s)_{III} - 1 - (P(f)_4 * P(f)_5)$$
$$= 1 - (.04 * .04) = 1 - .0016 = .998400$$

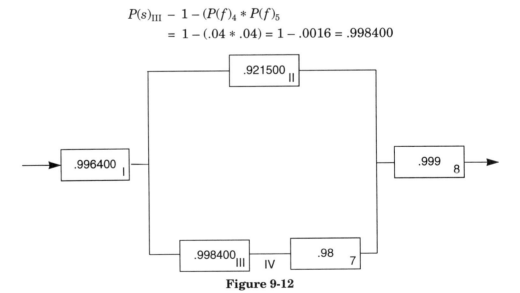

Figure 9-12

Solving for the probability of success for Part IV results in one block in parallel with the Block II. The probability of success for Comps 1 and 2, and Eng 2 is (Figure 9-13)

$$P(s)_{IV} = P(s)_{III} * P(s)_7 = .9984 * .98 = .9784320$$

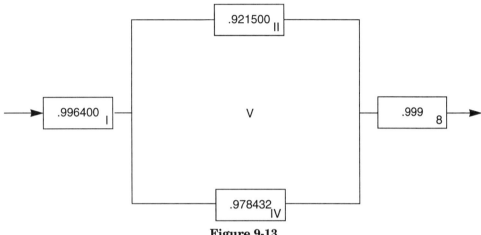

Figure 9-13

Solving for the probability of success for Part V results in one block in series with the rest of the system. The probability of success of Parts III and IV is (Figure 9-14)

$$
\begin{aligned}
P(s)_V &= 1 - (P(f)_{II} * P(f)_{IV}) \\
&= 1 - ((1 - .921500) * (1 - .9784320)) \\
&= 1 - (.0785000 * .0215680) \\
&= 1 - .001693088 = .998306912
\end{aligned}
$$

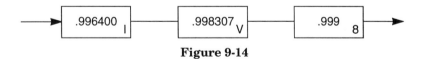

Figure 9-14

Solving for the probability of success for the final series gives the probability of success of the system:

$$
\begin{aligned}
P(s)_I &= P(s)_I * P(s)_V * P(s)_8 \\
&= .99640 * .998306912 * .99 = .993718294
\end{aligned}
$$

The probability of system failure is then simple to compute using the Complementary Law

$$P(f)_{\text{sys}} = 1 - P(s)_{\text{sys}} = 1 - .993718294 = .006281706$$

Let's examine a complex accident system. Figure 9-15 represents an accident system with the accompanying data for subsystem success and failure.

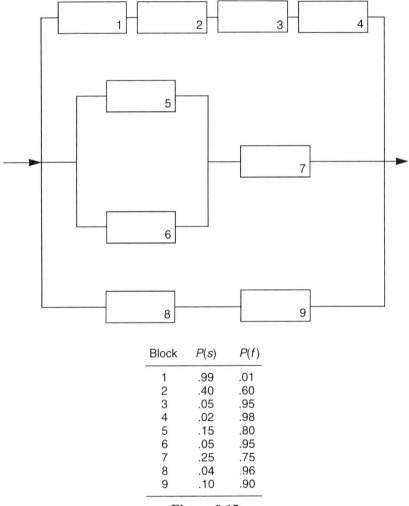

Block	P(s)	P(f)
1	.99	.01
2	.40	.60
3	.05	.95
4	.02	.98
5	.15	.80
6	.05	.95
7	.25	.75
8	.04	.96
9	.10	.90

Figure 9-15

Assume the accident system represents the ways that a company can experience a loss. Blocks 1 through 4 represent the fire chain, with block 1 as oxygen, block 2 as fuel, block 3 as ignition source, and block 4 as the chemical reaction. Blocks 5 through 7 represent injuries from ladder accidents, with block 5 as falls from the ladder, block 6 as ladders slipping, and block 7 the probability of an injury given the fall. Blocks 8 and 9 represent injuries due to slips and falls, with block 8 as a slip or fall and block 9 as an injury given a fall. What is the probability of a loss in this company?

The solution could be obtained by a formula, but it is easier to solve it one step at a time. Figure 9-16 shows the system divided into simple series/parallel systems.

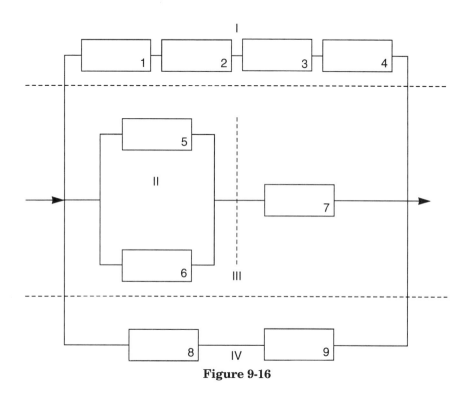

Figure 9-16

Solving for the probability of success of blocks 1 through 4 in series results in Part I in parallel with the rest of the system. The probability of having a fire $(P(s)_I)$ is

$$P(s)_I = P(s)_1 * P(s)_2 * P(s)_3 * P(s)_4$$
$$= .99 * .4 * .05 * .02 = .0003960$$

We now have the system of Figure 9-17. Solving for the probability of success of blocks 5 and 6 in parallel results in Part II in series with block 7 (Figure 9-18). The probability of having a ladder accident is

$$P(s)_{II} = 1 - P(f)_5 * P(f)_6$$
$$= 1 - (.8 * .95) = 1 - .76 = .2400$$

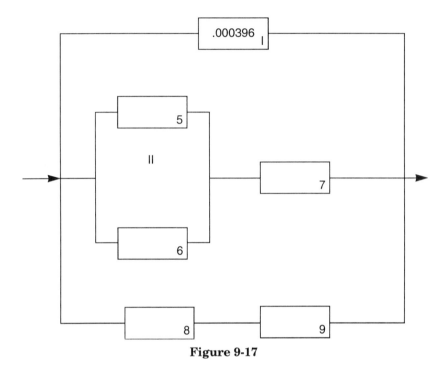

Figure 9-17

Solving for the probability of success of Part II and block 7 in series results in Part III in parallel with the rest of the system (Figure 9-19). The probability of having a loss due to a ladder accident is

$$P(s)_{\text{III}} = P(s)_{\text{II}} * P(s)_7 = .24 * .25 = .0600$$

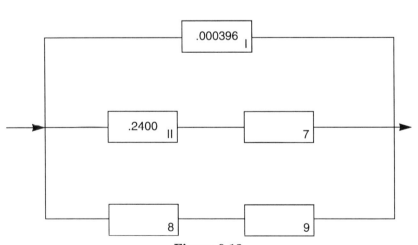

Figure 9-18

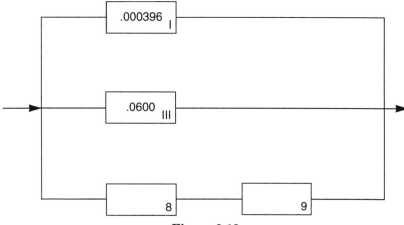

Figure 9-19

Solving for the probability of success for blocks 8 and 9 in series results in Part IV in parallel with the rest of the system (Figure 9-20). The probability of an injury due to falling is

$$P(s)_{\text{IV}} = P(s)_8 * P(s)_9 = .04 * .10 = .0040$$

Now, Parts I, III, and IV are in parallel. Therefore, the probability of success of the system (a loss) is

$$
\begin{aligned}
P(\text{loss})_{\text{sys}} &= 1 - (P(f)_{\text{I}} * P(f)_{\text{III}} * P(f)_{\text{IV}} \\
&= 1 - (1 - .000396) * (1 - .0600) * (1 - .004)) \\
&= 1 - (.999604 * .6400 * .9960) \\
&= 1 - .93586924896 \\
&= .06413075104
\end{aligned}
$$

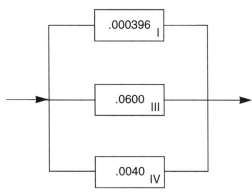

Figure 9-20

Accident systems can sometimes cause problems due to notations. Since the success of an accident system is an accident, various texts use different notations to indicate success and failure of the subsystems. For example, some texts present a number representing the mean time between failure (MTBF) in the system drawing. If this is the format presented, then the probability of success for a reliability system is represented by $P(0)$, and the probability of success for an accident system is represented by $P(f)$, where $P(0)$ and $P(f)$ are derived using the Poisson formula. Figure 9-21 will assist you in solving problems with this notation. The key is to develop your own table and use it to answer any question.

If Figure 9-21 represents a reliability system, $P(s)_1$ is equal to $P(0)$ derived by

$$P(0)_1 = e^{-t/m} = e^{-100/2.6E4} = e^{-3.84615E\text{-}3} = .996161233$$

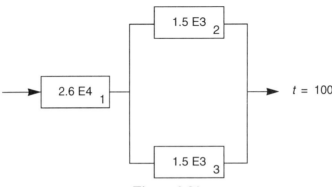

Figure 9-21

The table for this system is in Table 9-4.
The probability of success for Figure 9-21 as a reliability system is

$$
\begin{aligned}
P(s)_{\text{sys}} &= P(s)_1 * (1 - (P(f)_2 * P(f)_3)) \\
&= .996161233 * (1 - .064493015 * .064493015) \\
&= .996161233 * (1 - .00415934898) \\
&= .996161233 * .9958406510 \\
&= .99207851
\end{aligned}
$$

TABLE 9-4 BLOCK PROBABILITIES FOR FIGURE 9-21

Block	$P(s)$	$P(f)$
1	.996161233	.003838767
2	.935506985	.064493015
3	.935506985	.064493015

TABLE 9-5. BLOCK PROBABILITIES FOR
FIGURE 9-21 AS ACCIDENT SYSTEM

Block	$P(s)$	$P(f)$
1	.003838767	.996161233
2	.064493015	.93550698
3	.064493015	.93550698

If Figure 9-21 were an accident system with the same information provided in the blocks, then the table is given in Table 9-5.

The probability of success for Figure 9-21 as an accident system is

$$
\begin{aligned}
P(s)_{\text{sys}} &= P(s)_1 * (1 - (P(f)_2 * P(f)_3)) \\
&= .003838767 * (1 - .93550698 * .93550698) \\
&= .003838767 * (1 - .87517331) \\
&= .003838767 * .12482669 \\
&= 4.7918058 \text{ E-4}
\end{aligned}
$$

Once more, the key is to ensure that you develop your own table using the correct numbers for the probability of success and failure of the type system with which you are involved. Solve the following sample problems.

SAMPLE PROBLEMS

The following reliability system applies to Problems 9.1 through 9.6.

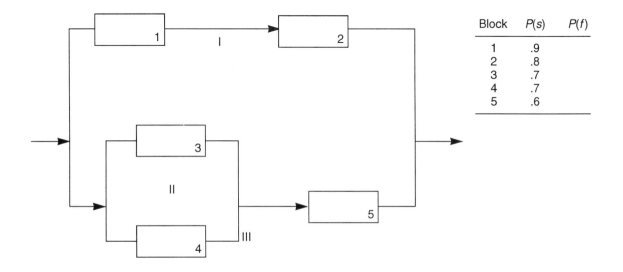

Block	P(s)	P(f)
1	.9	
2	.8	
3	.7	
4	.7	
5	.6	

9.1. What is the probability of failure for block 1?

    a. .4   (202)
    b. .3   (205)
    c. .2   (210)
    d. .1   (215)

9.2. What is the probability of success for Part I?

    a. .02   (220)
    b. .18   (225)
    c. .72   (260)
    d. .08   (267)

9.3. What is the probability of success for Part II?

    a. .09   (270)
    b. .91   (275)
    c. .49   (277)
    d. .51   (230)

9.4. What is the probability of success for Part III?

    a. .546   (237)
    b. .364   (242)
    c. .454   (246)
    d. .036   (253)

9.5. What is the reliability of the system?

    a. .12712   (166)
    b. .39312   (197)
    c. .60688   (180)
    d. .87288   (185)

9.6. What is the probability of failure of the system?

    a. .1271   (189)
    b. .1272   (193)
    c. 1.271 E-2   (175)
    d. 1.272 E-2   (200)

The following accident system applies to Problems 9.7 through 9.11.

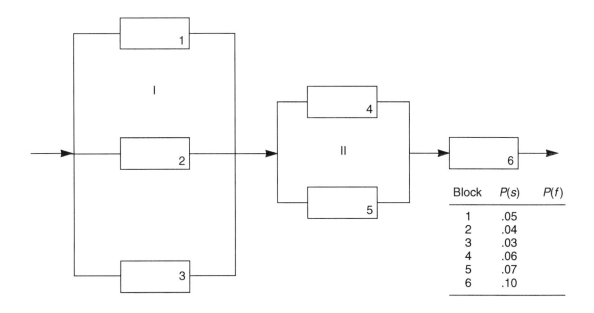

Block	P(s)	P(f)
1	.05	
2	.04	
3	.03	
4	.06	
5	.07	
6	.10	

9.7. What is the probability of failure for block 3?

    a. .9   (198)
    b. .93   (195)
    c. .97   (169)
    d. .95   (187)

9.8. What is the probability of success for Part I?

    a. 6.0000 E-5   (184)
    b. .88464   (179)
    c. 9.9994 E-1   (176)
    d. .11536   (167)

9.9. What is the probability of success for Part II?

    a. .1258   (256)
    b. .8742   (254)
    c. .0558   (249)
    d. .0658   (245)

9.10. What is the probability of an accident?

    a. .6960   (241)
    b. .001451   (238)
    c. .998549   (235)
    d. .3040   (181)

9.11. What is the probability of no accident?

      a. .997433   (276)
      b. .998549   (270)
      c. .6960   (268)
      d. .6864   (265)

9.12. Given the following block of an accident system, what is the probability of success (accident)?

$$m = 2.3E5 \qquad t = 200$$

      a. 9.991308 E-1   (259)
      b. 8.6919 E-4   (245)
      c. 9.995653 E-1   (223)
      d. 4.3469 E-4   (219)

9.13. Given the following block of a reliability system, what is the probability of success?

$$m = 1.2E3 \qquad t = 50$$

      a. 9.95842 E-1   (216)
      b. 4.1580 E-3   (211)
      c. 9.5919 E-1   (208)
      d. 4.0810 E-2   (205)

9.14. Given the following block of an accident system, what is the probability of failure (no accident)?

$$m = 1.4E2 \qquad t = 75$$

      a. 5.8525 E-1   (201)
      b. 9.4784 E-1   (232)
      c. 5.2162 E-2   (238)
      d. 4.1475 E-1   (243)

9.15. Given the following block of a reliability system, what is the probability of failure?

$$\boxed{m = 4.5\text{E}3} \qquad t = 80$$

a. 9.8238 E-1   (251)
b. 1.7621 E-1   (257)
c. 9.8238 E-2   (173)
d. 1.7621 E-2   (180)

10

Cut-Set Method

Assume Figure 10-1 is a reliability system. If the probability of success of each block is .9, what is the probability of success of the system? It appears that we might compute the probability as in Chapter 9 on event systems. However, this diagram cannot be dissected into purely parallel and series networks. Blocks 1 and 2 are in series, and blocks 1 and 3 appear to be in series, but block 3 is also in parallel with the combination of blocks 1 and 2. No matter how hard you try, there is no way to establish a true series/parallel relationship among all of the blocks. To solve such a problem, we need a method called cut sets.

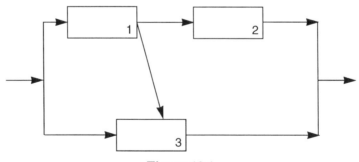

Figure 10-1

The concept behind cut sets is based on simply cutting the system. In other words, which blocks must fail at the same time in order to make the system fail? The system could fail in several ways. If blocks 1, 2, and 3 fail simultaneously, the system will fail. But simultaneous failures in blocks 1 and 2 will not cause system failure because the signal can pass through block 3. However, failure in blocks 1 and 3 or in blocks 2 and 3 will cause system failure. Each way that the system can fail is called a cut set. Therefore, the cut sets for Figure 10-1 are

$$1,2,3$$
$$1,3$$
$$2,3$$

In cut set 1,2,3, is it necessary for block 1 to fail in order for the system to fail? No. If blocks 2 and 3 fail, the system fails. The same is true for block 2 in the cut set 1, 2, 3. Since it is not necessary that all of the blocks in the cut set 1,2,3 fail in order to achieve system failure, the cut set 1,2,3 is not a minimum cut set. We are interested only in minimum cut sets. Therefore,

$$1,3$$
$$2,3$$

are the minimum cut sets. Now what do we do with them? The cut-set method is not as precise as the series/parallel method of solving networks. It only provides an upper bound on the probability of system failure. This means that once the cut-set solution is found, that solution is the highest probability possible for system failure. No matter what happens, the probability of system failure is less than the probability found from using cut sets.

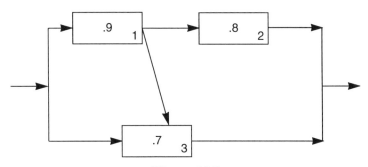

Figure 10-2

Let's use Figure 10-2 as an example of how to determine the probability of failure using cut sets. The numbers in the blocks represent the probability of success (reliability). The minimum cut sets have been determined to be 1,3 and 2,3. What is the probability that blocks 1 and 3 fail at the same time? From the Multiplication Law the probability of blocks 1 and 3 failing simultaneously is

$$P(f)_{1,3} = P(f)_1 * P(f)_3 = .1 * .3 = .03$$

(remember $P(f)_1 = 1 - P(s)_1 = 1 - .9 = .1$). The Multiplication Law also applies for the probability of 2,3 occurring:

$$P(f)_{2,3} = P(f)_2 * P(f)_3 = .2 * .3 = .06$$

If the system can fail either of two ways (1,3 or 2,3), the Addition Law takes effect and the probability of system failure is $P(1,3) + P(2,3) = .09$. Thus, no matter what happens, the probability of system failure is no greater than .09.

Let's see how this relates to the series/parallel method. In Figure 10-3 if we use the cut-set method, we obtain the same probability of failure as in the previ-

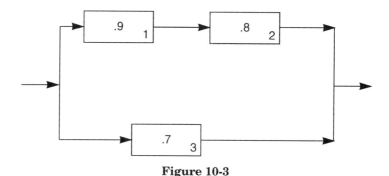

Figure 10-3

ous example (min cut sets = 1,3 and 2,3 (.09)). Now solving it by the series/parallel method gives

$$P(s)_I = P(0)_1 * P(0)_2 = .9 * .8 = .72$$
$$P(\text{suc})_{\text{sys}} = 1 - P(f)_I * P(f)_3 = 1 - ((1 - .72) * .3)$$
$$= 1 - (.28 * .3) = 1 - .084 = .91600$$
$$P(f)_{\text{sys}} = 1 - .91600 = .0840$$

This answer is very close to .09. But why didn't we get exactly .09? The reason is that .09 is an upper bound, whereas .084 is a precise number because series/parallel networks allow precise computations.

Now let's return to the problem of establishing minimum cut sets. Examine Figure 10-4 and determine the minimum cut sets. The first step is to list all cut sets we might want to consider. For this example, some cut sets might obviously not be minimum cut sets. However, there is no harm in listing additional cut sets, so we may list some that are not minimum cut sets while not listing others that are obviously not minimum cut sets.

To develop a systematic way of checking for cut sets, we suggest starting with block 1 and working our way through numerically. Thus, if block 1 fails, what else would have to fail in order to cut the system? Blocks 1 and 2 alone would not cut the system. Neither would 1,2,3 nor 1,2,3,4. The first cut set we might want to consider is 1,2,3,4,5. Hopefully it is obvious that nothing else needs to be added to this set in order to cut the system, so we will not list 1,2,3,4,5,6, even though it

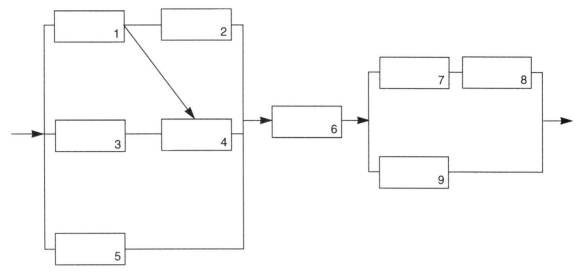

Figure 10-4

is a cut set. Also, can you see that in this section, block 5 must be in every cut set because the signal can always get to block 6 regardless of any combination of block failures apart from block 5.

Now we assume that 4 works. The next cut set numerically is 1,2,3,5. Next we assume that 3 and 4 work. No other combination of 1,2, and 5 cuts the system. Therefore, we skip 2 and examine the combinations with 1 and 3. The first is 1,3,4,5. Once more we examine for possible cut sets if 4 works. Won't 1,3,5 still cut the system? We now drop 3 from the combinations with 1. The only set left with block 1 is 1,4,5.

Now we assume 1 works and we examine cut sets beginning with 2. One advantage of proceeding numerically is that once we have passed a number as the starting number in a cut set, we need no longer consider it because all possible combinations involving that block have been examined. Therefore, the first cut set with 2 is 2,3,4,5. Now let's assume 4 works. We might want to list 2,3,5, but, on closer examination, we see that 2,3,5 is not a cut set because the system can work from 1 to 4 to 6, and so on. Once more we eliminate 3 from consideration and see that 2,4,5 is a cut set.

Now we look at combinations beginning with 3. The first is 3,4,5,6, because without 6 failing the system could go from 1 to 2 to 6, and so on. It should be obvious that 6 alone cuts the system, so any cut set with 6 in it (other than 6 by itself) is not a minimum cut set. Hence, we can skip combinations starting with 3,4, or 5.

After noting 6 is a cut set by itself, the next set we consider is 7,8,9. The remaining sets are 7,9 and 8,9. In summary, all of the cut sets we noted are

1,2,3,4,5	1,4,5	6
1,2,3,5	2,3,4,5	7,8,9
1,3,4,5	2,4,5	7,9
1,3,5	3,4,5,6	8,9

Now the challenge is to determine the minimum cut sets. The easiest way is to look for the set with the fewest elements, in this case 6. Now examine each cut set. If any contain 6, we do not need that cut set because it is not a minimum cut set (6 alone cuts the system). The only set listed with 6 in it is 3,4,5,6. We eliminate that set because 3, 4, or 5 need not fail for the system to fail as long as 6 alone fails.

Now we want the next set with the next fewest elements, in this example 7,9. We examine the sets to see if 7,9 is contained in any of them. We find the set 7,8,9 and eliminate it. The next set containing the fewest elements is 8,9. Since no other sets contain 8,9, we go to the next fewest elements, three (1,3,5). We can eliminate sets 1,2,3,4,5 and 1,3,4,5.

Continuing the process, we look for sets containing 1,4,5. Since 1,2,3,4,5 and 1,3,4,5 have already been eliminated, no other sets contain 1,4,5. The last set to examine is 2,4,5, which eliminates 2,3,4,5, leaving us with the minimum cut sets

$$
\begin{array}{ccc}
1,3,5 & 2,4,5 & 7,9 \\
1,4,5 & 6 & 8,9
\end{array}
$$

Figure 10-5 represents the same reliability network with the values for system success in each block. Based on the minimum cut sets we just found, what is the upper-bound probability of system failure?

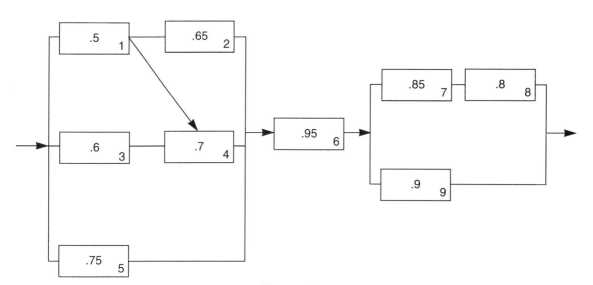

Figure 10-5

The answer is derived by determining the probability of each cut set and adding them.

$$
\begin{aligned}
P(f)_{1,3,5} &= .5 * .4 * .25 = .05 \\
P(f)_{1,4,5} &= .5 * .3 * .25 = .0375 \\
P(f)_{2,4,5} &= .35 * .3 * .25 = .02625
\end{aligned}
$$

$$P(f)_6 = .05 = .05$$
$$P(f)_{7,9} = .15 * .1 = .015$$
$$P(f)_{8,9} = .2 * .1 = .02$$
$$P(f)_{sys} = \Sigma P(f)_{\text{min cut sets}} = \overline{.19875}$$

No matter what happens, the probability of failure is less than .19875.

This process can be tedious, but with practice, much of the work can be done mentally. For example, we didn't need to list 1,2,3,4,5 or 1,2,3,5 as cut sets because we could see that they were not minimum cut sets.

SAMPLE PROBLEMS

The figure is for all of the sample problems.

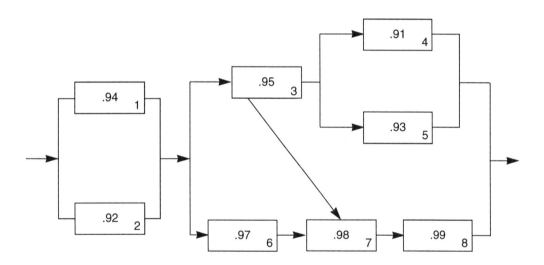

10.1. Which of the following is a cut set?

a. 1,6	(181)
b. 2,3	(187)
c. 4,5,6	(192)
d. 4,5,7	(276)

10.2. Which of the following is *not* a cut set?

a. 1,2,3	(199)
b. 3,6,7	(196)
c. 1,5,7	(195)
d. 4,5,8	(190)

10.3. Which of the following is a minimum cut set?

    a. 1,3,7   (185)
    b. 3,6   (183)
    c. 3,5,7   (178)
    d. 4,8   (170)

10.4. Which of the following is *not* a minimum cut set?

    a. 1,2   (257)
    b. 4,5,8   (254)
    c. 3,4,6   (250)
    d. 3,7   (246)

10.5. In which of the following are all sets true cut sets? (This does not mean that all of the cut sets are noted, but that all of the sets in the group are cut sets (not necessarily minimum cut sets).)

    a. 1,2   3,6   4,5,6   4,5,7   (242)
    b. 1,4,7   3,7   4,5,7   3,6,8   (239)
    c. 1,3,8   1,4,5,7   2,3,6,7   3,4,5,7   (235)
    d. 1,3,7   2,4,5,6   2,4,5,7   3,4,5,6   (204)

10.6. In which of the following are all sets minimum cut sets? (This does not mean that all minimum cut sets are noted, but that all of the sets in the group are minimum cut sets.)

    a. 1,2   3,7   4,5,7   4,5,8   (209)
    b. 1,2   3,6   4,5,7   5,7,8   (212)
    c. 3,6   3,7,8   4,5,7   4,5,8   (217)
    d. 3,6   3,7   3,8   3,4,6   4,5,8   (223)

10.7. Which of the following contains all of the minimum cut sets? (*All* of the sets may not be minimum, but *all* of the sets are cut sets and *all* of the minimum cut sets are in the correct answer.)

    a. 1,2   3,4,7   3,6   3,7,8   3,8   4,5,7   4,5,8   (259)
    b. 1,2   3,4,8   3,6   3,7   3,8   4,5,6,7   4,5,7   4,5,8   (264)
    c. 1,2   2,3,6   3,6,7   3,7   3,8   4,5,7   4,5,8   (269)
    d. 1,2   2,3,7   3,6   3,7   3,8   4,5,6   4,5,7   4,5,7,8   (274)

10.8. Which of the following is the correct list of minimum cut sets?

    a. 1,2   3,6   3,7   4,5,7   4,5,8   (279)
    b. 1,2   3,6   3,7   3,8   4,5,7   4,5,8   (272)
    c. 1,2   3,6   3,7   4,5,6   4,5,7   4,5,8   (169)
    d. 1,2   3,4,5   3,6   3,7   3,8   4,5,7   (266)

10.9. What is the upper bound on the probability of system failure?

    a. .0076780   (263)

    b. .0074890   (228)

    c. .0082410   (224)

    d. .0079890   (221)

11

Boolean Algebra

Entire college courses are taught on Boolean algebra, so an extensive discussion of the logic behind Boolean algebra is beyond the scope of this guide. However, the principles and concepts of Boolean algebra can be very helpful in occupational safety and health. This guide defines some key terms and symbols of Boolean algebra and shows how to use Venn diagrams to solve Boolean algebra problems.

Boolean algebra is a system that investigates the relationship between variables. The variables can be logic statements or categories of systems. This discussion deals primarily with categories of systems, but the rules apply to both.

The first symbol is for a category of a system. In the rest of this guide, these categories are called groups. Groups are represented by letters. For example, if one group is males, then males could be represented by the letter M. Every group in a system has a probability of occurrence. Thus, the probability of someone in the system being a male is written $P(M)$. The probability of not being male is written $P(\overline{M})$ or $P(M')$. We use M' in this guide.

The next concept is the universe, w. The universe is represented by a rectangular box as in Figure 11-1. The universe represents the entire system of interest, or everything over which you have control. If something falls outside your system or area of control, it is considered nonexistent and has no probability of occurrence. Therefore, everything that can occur within the universe must have a probability of unity: $P(w) = 1$ and $P(w') = 0$.

The logic symbols for union and intersection will now be discussed. Union means that either of two groups being considered satisfies the criteria. The symbol for union is $\cup$: $P(A \cup B)$ then means the probability of A occurring or B occurring. Because the probability is either/or, the Addition Law applies and the

Figure 11-1

probabilities of the individual groups can be added. However, other factors must be considered.

One is the issue of mutual exclusivity. Mutual exclusivity questions if members of group A can also be members of group B. As an example, if A is women over 30 and B is women under 25, then A and B are mutually exclusive, because a female cannot be over 30 and under 25 at the same time. If, however, groups are not mutually exclusive, then some members of one group might be members of the other group. If the groups are not mutually exclusive and the Addition Law is used, then part of the group is added twice. (This point is easily seen when we discuss Venn diagrams.) Therefore the proper calculation of the union is

$$P(A \cup B) = P(A) + P(B) - P(A \cap B)$$

where $\cap$ means intersection. Intersection is another way of combining groups. Intersection means the probability of an event belonging to both groups. There are three ways to solve for $P(X \cap Y)$. The first is the simplest. If X and Y are mutually exclusive, then $P(X \cap Y) = 0$. If the two groups are not mutually exclusive and intersect, then, generally, the Multiplication Law can be used and

$$P(X \cap Y) = P(X) * P(Y)$$

Another factor affects the computation of intersections, namely the question of dependence versus independence. The preceding formula is true if the groups are independent. Groups are independent if one has no effect on the other. For example, we have no reason to assume that the color of clothes we wear influences the number of colds we catch. Therefore, the two groups, color of clothing and colds, are independent of each other. If groups are independent, then

$$P(X \cap Y) = P(X) * P(Y)$$

However, sometimes groups are dependent. For example, it is logical to assume that in order to drown during work, a person must work around some volume of fluid. Therefore, if group D is occupational drownings and group N is workers who work around large bodies of fluid, a dependency exists between the groups.

When groups are dependent, we need to have enough information to determine the strength of the dependency. In industry this information should be available. Since this relationship (dependency) is so variable, for this guide, dependency will require that you be told the value of $P(D \cap N)$.

This discussion leads to Venn diagrams. Groups are represented by circles. The circles are not drawn to scale and are not true representations of the size of the groups. However, the relationship between the placement of two circles is meaningful.

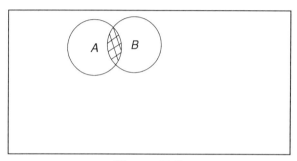

Figure 11-2

If two groups are not mutually exclusive, the circles intersect (overlap) one another. Figure 11-2 shows the intersection of A and B. Notice that if we take the probability of the union of A and B ($P(A \cup B)$) and simply add the probabilities, the shaded area is counted twice. Thus, we must subtract the intersection of A and B ($P(A \cap B)$). If the groups are mutually exclusive, the circles do not touch. There is no way to show dependency, however. We must note dependency through the rest of the information. In this guide, dependency is noted by providing the value for $P(A \cap B)$.

One more concept needs to be discussed: the unique relationships of the universe to any group. The following equations are always true (since Boolean algebra is transitive, the transitives are also true):

$$P(w \cup R) = 1$$
$$P(w \cap R) = P(R)$$

In other words, the probability of a union between the universe and a group is always 1, and the probability of an intersection between the universe and any group is the probability of that group.

Many laws of regular algebra apply to Boolean algebra. However, rather than attempting to memorize the rules, there is an easier way to solve many of the problems—simply use Venn diagrams. At this point, the development of a Venn diagram may be the best way to demonstrate the above principles.

Let's construct the Venn diagram for the following example. Let group A represent people who smoke more than two packs of cigarettes a day. Then $P(A)$ is the probability of smoking more than two packs a day. This probability could be determined a posteriori by dividing the number of people in a sample by the number of those who smoke more than two packs a day. Assume $P(A) = .2$.

Now let's take another group, B, which is women between the ages of 20 and 29 inclusive. Based on the sample, the probability of belonging to this group is $P(B) = .25$.

One other group, C, is women between the ages of 30 and 39 (inclusive). Assume the probability is $P(C) = .3$.

It is logical to assume that members of groups B or C could belong to A, so A is not mutually exclusive from either B or C. However, a woman cannot be a member of B and C and the same time, so these groups are mutually exclusive. Therefore, the Venn diagram is as in Figure 11-3.

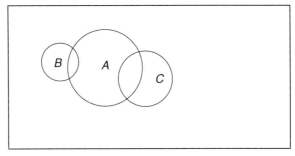

Figure 11-3

Now we must determine if the groups are dependent. In this case there is no logical reason to assume dependency, so we consider each of them to be independent. It's a good idea to note the group probabilities on the diagram, as in Figure 11-4.

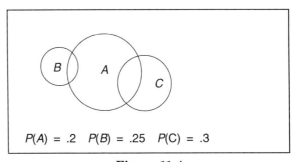

Figure 11-4

Now you are ready to examine some relationships.

Example

For Figure 11-4, determine $P(B \cup C)$

Solution   Since the two groups are mutually exclusive, no overlap must be subtracted. By the Addition Law,

$$P(B \cup C) = P(B) + P(C) = .25 + .3 = .55$$

The solution represents the probability that someone in the universe is either a female between the ages of 20 and 29 or a female between the ages of 30 and 39.

Example

For Figure 11-4, determine $P(A \cap B)$.

Solution   Intersection means you want the probability of belonging to both groups. Since the groups are independent but not mutually exclusive, $P(A \cap B)$ is the probability of A *and* the probability of B occurring at the same time. By the Multiplication Law,

$$P(A \cap B) = P(A) * P(B) = .2 * .25 = .05$$

This number, 0.5, is the probability of being a female between 20 and 29 who smokes more than 2 packs a day.

Example

For Figure 11-4, determine $P(A \cup C)$.

Solution   Since the two groups are not mutually exclusive, we must subtract the overlap. Since they are independent, the overlap is simply the mathematical intersection. By the Addition Law,

$$P(A \cup C) = P(A) + P(C) - P(A \cap C)$$
$$= .2 + .3 - (.2 * .3) = .5 - .06 = .44$$

The solution represents the probability that someone in the universe is either female between the ages of 30 and 39 or smokes more than 2 packs a day.

Example

For Figure 11-4 determine $P(C \cup w)$.

Solution   The probability of the union of anything with the universe is, by definition, 1.

Example

For Figure 11-4 determine $P(B \cup C) \cup P(A)$.

Solution This example involves another law that applies to Boolean algebra, the Associative Law. According to this law, the problem can be rewritten as

$$P(B \cup A) \cup P(C \cup A) = (P(B) + P(A) - P(B \cap A))$$
$$+ (P(C) + P(A) - P(C \cap A) - P(B \cup A) \cap P(C \cup A))$$

Since the new subgroups formed by $P(B \cup A) \cup P(C \cup A)$ are dependent because of A's relationship with itself, the mathematics can become rather sticky. This is when Venn diagrams become so useful. We suggest you use colors in your drawings to enhance comprehension.

As in regular algebra, we solve the probability in parentheses first (i.e., $P(B \cup C)$). Hence, we draw lines in everything we want (Figure 11-5). Since we want the union of B and C, we want either B or C—that is, everything with a line in it.

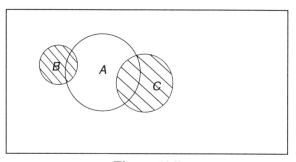

Figure 11-5

Next we draw the Venn diagram for the union of this group (everything with a line) and A (Figure 11-6). Again we want a union, so everything with a line in it is included in the answer. Now we are ready to put numbers in place. Simply add A, B, and C and subtract the areas counted twice—the areas where the lines cross, which are the intersections of A and B, and A and C.

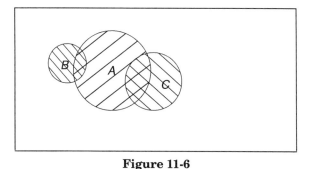

Figure 11-6

$$P(B \cup C) \cup P(A) = P(A) + P(B) + P(C) - P(A \cap B) - P(A \cap C)$$
$$= .2 + .25 + .3 - .05 - .06 = .6400$$

The solution represents the probability that someone in the universe belongs to at least one of the groups—female between the ages of 20 and 29, female between the ages of 30 or 39, or smoker of more than 2 packs a day.

Example

For Figure 11-4 determine $P(A \cup C) \cap P(B)$.

Solution   Algebraic laws could be applied, resulting in

$$P(A \cap B) \cup P(C \cap B)$$

However, the Venn diagram is easier. First draw lines in the area wanted for $P(A \cup C)$ (Figure 11-7). Since it is a union, we want the entire lined area.

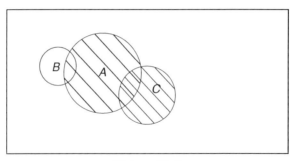

Figure 11-7

Next, we line $P(B)$ (Figure 11-8). Since we want the intersection of these two sets, we want the area where the lines cross. Figure 11-9 shows this area with lines. This area is simply $P(A \cap B)$.

$$P(A \cap B) = .2 * .25 = .05$$

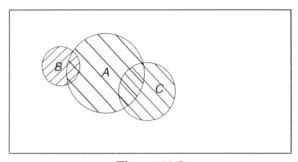

Figure 11-8

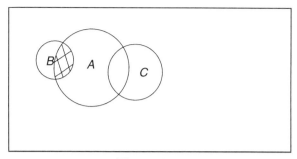

Figure 11-9

This answer means that we can expect 5% of the population to be females between 20 and 29 who smoke more than 2 packs a day.

Now let's add another group to our universe. Group D will represent infants (< one year old) who attend the day care facility in our plant. They represent .01 of the sample population. Since it is impossible to be a female between the ages of 20 and 39 and still be an infant, group D is obviously mutually exclusive from groups B and C. Likewise, infants do not smoke, so group D is mutually exclusive from group A. Figure 11-10 is the new Venn diagram.

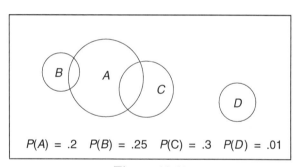

Figure 11-10

Let's add one more group. Group E will be teenagers (15–19) who perform summer jobs in our plant. There is a probability of .03 of being in group E. Obviously group E is mutually exclusive from groups B, C, and D. However, there is no reason to believe that some teenagers do not smoke two packs a day. Therefore, groups E and A intersect. However, there is one complication. Now assume it is illegal to purchase smoking material before the age of 18. Since teenagers from 15 to 17 years old work during the summer, there is a dependency between groups E and A. Because of the illegality, we should not expect $P(A \cap E) = .006$, but there is still a probability that some underage teenagers smoke more than two packs a day. In practice this could be determined from the raw data obtained in the measurement of the sample, but for this discussion assume we are given that $P(A \cap E) = .002$. Figure 11-11 represents all of the groups.

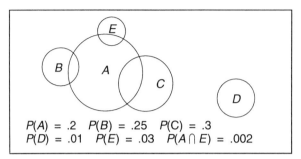

Figure 11-11

It is conceivable that the sum of all of the groups is greater than 1. How is this possible? Since a person can be a member of more than one group, he or she would be counted twice if we simply added the groups. Therefore, we must subtract any intersection (overlap) in the diagrams. As long as intersections are subtracted the total probability will never be greater than 1. We will use Figure 11-11 to work the following examples.

Example

Find $P(D' \cup E')$.

Solution   The easiest solution is always the Venn diagram. However, it is not necessary to draw the entire picture, so draw only what you need. Remember, the relative size makes no difference, but the groups must be in proper relationship (intersection or not).

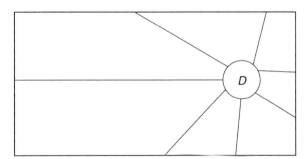

Figure 11-12

First, we draw D and determine what we need. The problem calls for "not D," so the lines start at the edge of D and extend to the edge of the universe (Figure 11-12). Thus, mathematically we have everything except $P(D)$. But we won't use any numbers yet. Instead we finish the Venn diagram.

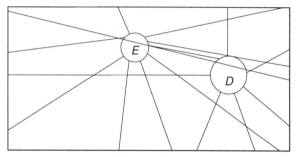

Figure 11-13

Putting E in the diagram and drawing lines starting at its edge and going to the border of the universe, we see that the lines cover the entire universe (Figure 11-13). Now since we want the union of D' and E', we want everything with a line in it. Therefore,

$$P(D' \cup E') = 1$$

Logically, this answer makes sense. The probability of not being an infant in day care or of not being a teenager working in the plant covers everyone, since someone who is not an infant or a teenager obviously belongs to one of the other groups in the universe, or someone who is not an infant could be a teenager, or someone who is not a teenager could be an infant. That is harder to put into words than it is to see in the picture.

Example

Find $P(D' \cap E')$

Solution   Look at the above solution to $P(D' \cup E')$. Note the Venn diagram is exactly the same as for this solution. But now we want the probability of the intersection—where the lines cross. In Figure 11-13 the lines cross everywhere except within groups D and E. Therefore, we want everything except D and E. Hence,

$$P(D' \cap E') = 1 - P(D) - P(E) = 1 - .01 - .03 = .96$$

This means that the probability of being everything except an infant or a teenager is .96.

Example

Find $P(B \cup E) \cap P(C' \cap A)$

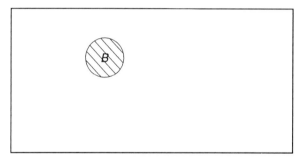

Figure 11-14

Solution   First find the area of interest in $P(B \cup E)$ by drawing it. First draw B (Figure 11-14).

Then we add E (Figure 11-15).

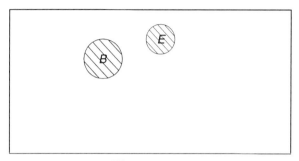

Figure 11-15

Since we want the union, we want everything with a line in it, so the entire picture remains.

Now for $P(C' \cap A)$ we draw a separate diagram (for clarity). See Figure 11-16.

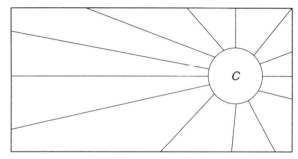

Figure 11-16

Then we add *A* (Figure 11-17).

We want the intersection, so we want the area where the lines cross (intersect). See Figure 11-18. The shaded area is the only area of interest.

Now we add this area to our first partial results (Figure 11-15) and get Figure 11-19.

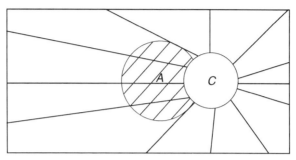

Figure 11-17

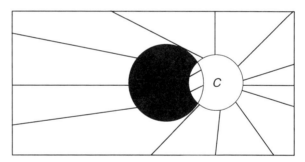

Figure 11-18

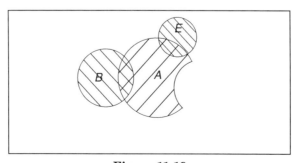

Figure 11-19

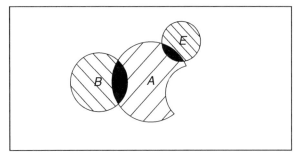

Figure 11-20

We want the intersection. The only areas where lines cross are shown in Figure 11-20.

Therefore,

$$P(B \cup E) \cap P(C' \cap A) = P(B) * P(A) + P(E) * P(A)$$
$$(.25 * .2) + (.03 * .2) = .05 + .006 = .056$$

This appears more complicated than it really is. Using colored pencils, or being very careful we could have obtained the solution with just one diagram (Figure 11-21). We went through the detailed steps to ensure that you understand the process.

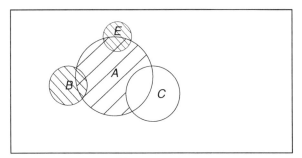

Figure 11-21

Now it is time to allow you to practice your skills.

SAMPLE PROBLEMS

Use the following Venn diagram and probabilities for all the sample problems.

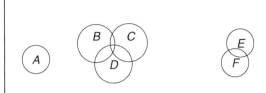

$P(A) = .11$ $P(B) = .25$ $P(C) = .1$ $P(D) = .05$

$P(E) = .15$ $P(F) = .2$ $P(E \cap F) = .06$

11.1. $P(A \cup E) =$

    a. .16   (231)
    b. 0   (237)
    c. .31   (240)
    d. .26   (244)

11.2. $P(B \cup D) =$

    a. .3   (250)
    b. .2875   (255)
    c. .0125   (166)
    d. .325   (173)

11.3. $P(D \cap F) =$

    a. .01   (178)
    b. .1   (184)
    c. 0   (189)
    d. .06   (215)

11.4. $P(C \cap D) =$

    a. .005   (198)
    b. .15   (200)
    c. .145   (196)
    d. .01   (193)

11.5. $P(E \cup F) =$

    a. .32   (188)
    b. .35   (165)
    c. .29   (181)
    d. .06   (177)

11.6. $P(A' \cup F') =$

    a. 1   (170)
    b. .69   (257)
    c. .31   (254)
    d. .25   (250)

11.7. $P(B' \cup C') =$

    a. .325   (245)
    b. .855   (241)
    c. .975   (238)
    d. .025   (233)

11.8. $P(C' \cap A) =$

    a. .11   (202)
    b. 1   (208)
    c. .89   (212)
    d. .85   (216)

11.9. $P(F' \cap B) =$

    a. .8   (222)
    b. .25   (226)
    c. .2   (263)
    d. .1   (269)

11.10. $P(C' \cap D') =$

    a. .995   (242)
    b. .755   (275)
    c. .855   (279)
    d. 0   (272)

11.11. $P(A \cup B) \cup E =$

    a. .51   (269)
    b. .41   (266)
    c. 0   (228)
    d. .15   (224)

11.12. $P(E \cap F) \cup B =$

    a. .25   (219)
    b. .275   (194)
    c. 0   (211)
    d. .31   (206)

11.13. $P(A \cap E) \cup C =$

    a. .11   (200)
    b. .1   (233)
    c. 0   (237)
    d. .2   (241)

11.14. $P(C \cap D) \cap F =$

    a. .205   (247)
    b. .215   (253)
    c. 0   (258)
    d. .1   (172)

11.15. $P(B' \cap C) \cup A =$

    a. .46   (179)
    b. .135   (185)
    c. 0   (280)
    d. .185   (193)

11.16. $P(E \cup C) \cup A' =$

    a. .89   (197)
    b. .25   (195)
    c. .36   (190)
    d. 1   (187)

11.17. $P(B \cap D) \cap F' =$

    a. .05   (179)
    b. .125   (174)
    c. .0125   (168)
    d. 1   (257)

11.18. $P(B \cap C) \cap D =$

    a. 1   (253)
    b. .00125   (249)
    c. .0125   (244)
    d. .125   (241)

11.19. $P(B \cup C) \cup D =$

    a. .4000   (236)
    b. .3625   (233)
    c. .34625   (202)
    d. .3600   (210)

11.20. $P(w \cap B) \cup E =$

    a. 1   (216)

    b. 0   (222)

    c. .4   (228)

    d. .15   (264)

11.21. $P(w \cup C) \cap D =$

    a. .005   (269)

    b. .1   (276)

    c. 1   (234)

    d. .05   (223)

12

Fault-Tree Analysis

Fault-tree analysis was developed in the early 1960s to provide both qualitative and quantitative analysis of a system. The term *fault tree* is a bit erroneous in that the same method can also be used to determine reliability. However, this discussion will be based on fault (accident) systems.

Numerous texts provide in-depth discussions of the development and analysis of fault trees. We assume that you have access to these texts and already have a basic knowledge of fault-tree analysis. This guide gives examples of basic quantitative analysis of a fault tree, with emphasis on the use of Boolean logic to simplify a fault tree. The end result is a Boolean equivalent tree which acts similarly to the cut-set method to provide an upper bound on the probability of an accident (failure) given the original fault tree.

The Boolean equivalent reduces a complex tree to the pertinent parts driving the system. As such, the Boolean equivalent is equivalent to the cut sets of the system. This concept will be developed in more detail later.

A review of the definitions of the basic symbols is all we need before solving the example problems. Remember that a rectangle represents a fault that needs analysis, a circle represents a basic fault—one requiring no further investigation—and faults are connected through a series of gates. The two primary gates are the OR gate and the AND gate. Their meaning is loosely summarized by the following mathematical equivalents:

$$\bigcap = \text{OR gate} = \cup = + \qquad \bigsqcup = \text{AND gate} = \cap = *$$

Fault-tree analysis is based on Boolean algebra. Therefore, addition (+) is based on the concept of union ($\cup$) and multiplication (*) is based on the concept of intersection ($\cap$). Hence, we cannot simply look at a fault tree and start adding and multiplying probabilities based on gate type. The key concept is remembering that the union (+) of anything with the universe (w) equals 1. This means that if we have a probability of $P(1 + P(A) + P(B) + \cdots + P(X))$, it does not matter what the individual probabilities are because the entire term will reduce to 1.

This leads to the second key concept. The intersection (*) of anything with the universe (w)—and by definition $w = 1$—is simply the anything being intersected. Therefore, $A * (1) = 1$ and $BC * (1 + B + C + \cdots + X) = BC$. With these two rules in mind, it becomes fairly easy to reduce complex fault trees to their Boolean equivalents. Once we have the Boolean equivalent, we substitute the individual probabilities and determine an upper bound on the probability of occurrence.

Example

Given Figure 12-1, what is the probability of the fault?

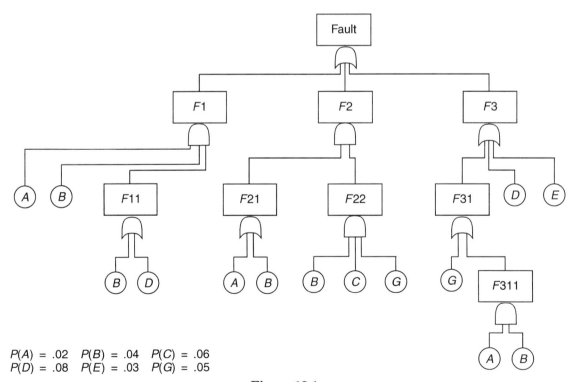

P(A) = .02 P(B) = .04 P(C) = .06
P(D) = .08 P(E) = .03 P(G) = .05

Figure 12-1

Solution   In Figure 12-1, the ultimate fault might be an injury from falling off a ladder. $F1$, $F2$, and $F3$ might then represent, respectively, the ladder slipping, the ladder blowing over, or human error. Each subfault can then be traced to its basic elements.

Figure 12-1 can be solved in numerous ways. However, to develop proficiency, we recommend that you start by working your way through the tree one level at a time. By one level, we mean those symbols on the same horizontal line with each other.

First, look at the top rectangle and use that as your dependent variable for this process. Therefore, the first equation is

$$\text{Fault} = F1 + F2 + F3$$

because the first gate is an OR gate and all three of these blocks are connected to it. Next, look at the block on the left (work top to bottom, left to right). Since there is an AND gate connecting $F1$ to its subordinate faults, write the equation

$$F1 = AB(F11)$$

(Note that the abbreviated form of indicating multiplication is used. AB is equal to $A * B$, and $AB(F11)$ is equivalent to $A * B * F11$.)

Next, write the equation for $F2$:

$$F2 = F21(F22)$$

The last equation on this level is for $F3$:

$$F3 = F31 + D + E$$

This logic continues until you have written the equation for every level. The equations are thus

$$
\begin{aligned}
F &= F1 + F2 + F3 \\
F1 &= AB(F11) \\
F2 &= F21(F22) \\
F3 &= F31 + D + E \\
F11 &= B + D \\
F21 &= A + B \\
F22 &= BCG \\
F31 &= G + F311 \\
F311 &= AB
\end{aligned}
$$

Once you have written the equations for every level, return to the first level and substitute the values for any term not at its basic level—it does not possess an assigned probability. After you become proficient at this process, you can combine many of the following steps. However, we substitute only one value at a time. The equations are as follows:

$$F = F1 + F2 + F3$$
$$F = AB(F11) + F2 + F3$$
$$F = AB(F11) + F21(F22) + F3$$
$$F = AB(F11) + F21(F22) + F31 + D + E$$
$$F = AB(B + D) + F21(F22) + F31 + D + E$$
$$F = AB(B + D) + (A + B)(F22) + F31 + D + E$$
$$F = AB(B + D) + (A + B)(BCG) + F31 + D + E$$
$$F = AB(B + D) + (A + B)(BCG) + G + F311 + D + E$$
$$F = AB(B + D) + (A + B)(BCG) + G + AB + D + E$$

After all of the basic faults are substituted into the equation, we begin reducing the terms. To simplify this process, we expand the equation to eliminate complex terms. The first expansion is $AB(B + D)$:

$$F = AB + BD + (A + B)(BCG) + G + AB + D + E$$

Remember that the intersection of anything with itself is still itself. That is why $AB(B + D)$ becomes $AB + BD$. The next step gives

$$F = AB + BD + ABCG + BCG + G + AB + D + E$$

Now that the equation is in its simplest terms (no compound terms), we reduce it by factoring. There are many ways to factor the equation. Each method should give the same answer. It is recommended that you follow the method here to ensure that all terms are considered in the easiest way.

In the cut-set method, to determine the final solution after you have all the minimum cut sets, you add the cut sets. The foregoing equation is very similar to that method in that each of the groups separated by a "+" represents a cut set. First we examine the equation from left to right, looking for the lowest number of terms in a cut set. In this equation, the first, smallest cut set is G. Factoring all of the terms containing G gives

$$F = AB + BD + \cancel{ABCG} + \cancel{BCG} + \cancel{G} + AB + D + E \quad \text{(original)}$$
$$F = G(ABC + BC + 1) + AB + BD + AB + D + E \quad \text{(factored)}$$

As you factor each term, put a line through it because that term need not be considered again. The next cut set with the fewest terms is D. Factoring D from the other terms yields

$$F = AB + \cancel{BD} + \cancel{ABCG} + \cancel{BCG} + \cancel{G} + AB + \cancel{D} + E \quad \text{(original)}$$
$$F = G(ABC + BC + 1) + D(B + 1) + AB + AB + E \quad \text{(factored)}$$

Continuing, we have

$$F = AB + \cancel{BD} + \cancel{ABCG} + \cancel{BCG} + \cancel{G} + AB + \cancel{D} + \cancel{E} \quad \text{(original)}$$
$$F = G(ABC + BC + 1) + D(B + 1) + E + AB + AB \quad \text{(factored)}$$

Notice that we did not really do anything because there were no terms from which to factor an E. Now that all single terms are exhausted, we turn our attention to the double-letter terms. There is only one, so we have

$$F = \cancel{AB} + \cancel{BD} + \cancel{ABCG} + \cancel{BCG} + \cancel{G} + \cancel{AB} + \cancel{D} + \cancel{E} \quad \text{(original)}$$
$$F = G(ABC + BC + 1) + D(B + 1) + E + AB(1 + 1) \quad \text{(factored)}$$

If there were three-letter terms, we would now factor them. There are none, so we are ready for the next step. Factoring allowed us to put the universe (1) in union with other terms. Hence, those terms are unnecessary because the unions reduce to 1. Therefore, the equation becomes

$$F = G + D + E + AB$$

Rearranging in alphabetical order leaves

$$F = AB + D + E + G$$

This equation is the Boolean equivalent of the original fault tree. In essence, these terms act like minimum cut sets (AB, D, E, G). Therefore, the probability of the fault is

$$F = AB + D + E + G = .02(.04) + .08 + .03 + .05 = .1608$$

and the tree can be redrawn as in Figure 12-2.

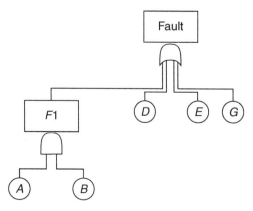

Figure 12-2

In some problems, only the equation, not the fault tree, may be given, and you must determine its Boolean equivalent. It is only necessary then to use the Boolean logic to reduce the equation.

Example

Find the Boolean equivalent of

$$F = AB + D + EGJ + AC + B + ACEH + DGJ + AEGHJ$$

Solution

$$F = D(1 + GJ) + AB + EGJ + AC + B + ACEH + AEGHJ$$
$$F = D + B(A + 1) + EGJ + AC + ACEH + AEGHJ$$
$$F = D + B + AC(1 + EH) + EGJ + AEGHJ$$
$$F = D + B + AC + EGJ(1 + AH)$$
$$F = B + D + AC + EGJ$$

If desired, we could draw the Boolean equivalent tree from this equation.

If the fault tree in the first example represents what leads to the fault, then what is required to prevent the fault from occurring? We can find out by developing the path sets. Path sets represent the opposite of cut sets and can be determined from the dual. To draw the dual, we change OR gates to AND gates and AND gates to OR gates. Figure 12-3 is the dual of Figure 12-1.

Note that in the dual all events appear primed (A'), which signifies that the event does not occur. Some texts use a bar over the event $(\overline{A})$. To determine the path sets, we must find the Boolean equivalent of this dual fault tree. The process is the same as for developing the Boolean equivalent of the original fault tree—work our way down, level by level. Solving for the path sets results in the following equations:

$$\text{Fault}' = F1'(F2')(F3')$$
$$F1' \quad = A' + B' + F11'$$
$$F2' \quad = F21' + F22'$$
$$F3' \quad = F31' \, D'E'$$
$$F11' \quad = B'D'$$
$$F21' \quad = A'B'$$
$$F22' \quad = B' + C' + G'$$
$$F31' \quad = G'(F311')$$
$$F311' = A' + D'$$

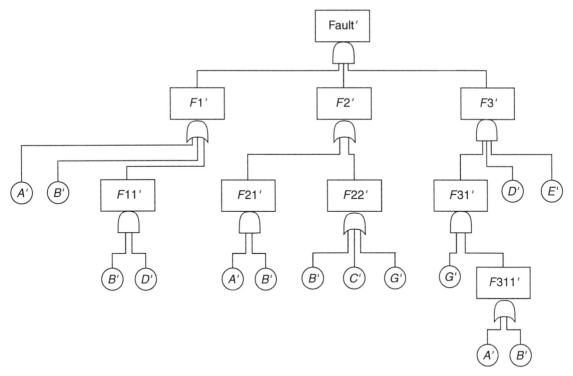

Figure 12-3

Substituting the values into each level yields

$$\text{Fault}' = (A' + B' + F11')F2'F3'$$
$$F' = (A' + B' + F11')(F21' + F22')F3'$$
$$F' = (A' + B' + F11')(F21' + F22')(F31' \, D'E')$$
$$F' = (A' + B' + B'D')(F21' + F22')(F31' \, D'E')$$
$$F' = (A' + B' + B'D')(A'B' + F22')(F31' \, D'E')$$
$$F' = (A' + B' + B'D')(A'B' + B' + C' + G')(F31' \, D'E')$$
$$F' = (A' + B' + B'D')(A'B' + B' + C' + G')(G'(F311')D'E')$$
$$F' = (A' + B' + B'D')(A'B' + B' + C' + G')(G'(A' + D')D'E')$$
$$F' = (A' + B' + B'D')(A'B' + B' + C' + G')(A'D'E'G' + D'E'G')$$

Because of all the multiplications, these terms get lengthy rapidly. Rather than write such long terms, reduce whenever possible. Therefore, the previous equation becomes

$$F' = (A' + B'(1 + D'))(A'B' + B' + C' + G')(A'D'E'G' + D'E'G')$$
$$F' = (A' + B')(B'(A' + 1) + C' + G')(A'D'E'G' + D'E'G')$$
$$F' = (A' + B')(B' + C' + G')(D'E'G'(A' + 1))$$
$$F' = (A' + B')(B' + C' + G')(D'E'G')$$

Combining terms leads to

$$F' = (A'B' + A'C' + A'G' + B' + B'C' + B'G')(D'E'G')$$

This equation is the multiplication of the terms $(A' + B')$ and $(B' + C' + G')$. Continuing, by first reducing where possible, leads to

$$F' = (B'(A' + 1 + C' + G') + A'C' + A'G')(D'E'G')$$
$$F' = (B' + A'C' + A'G')(D'E'G')$$
$$F' = B'D'E'G' + A'C'D'E'G' + A'D'E'G'$$
$$F' = B'D'E'G' + A'D'E'G'(C' + 1)$$
$$F' = B'D'E'G' + A'D'E'G'$$
$$F' = A'D'E'G' + B'D'E'G'$$

The last equation represents the path sets $A'D'E'G'$ and $B'D'E'G'$ and can be drawn as in Figure 12-4.

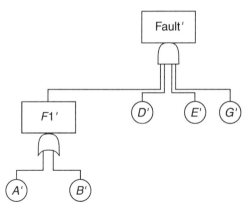

Figure 12-4

Notice that the Boolean equivalent of the dual is the dual of the Boolean equivalent of the original fault tree. If you are confident of the solution for the Boolean equivalent of the original fault tree, to determine the path sets you need not draw the dual of the original fault tree. Merely draw the dual of the Boolean equivalent of the original fault tree. Once done, it is sometimes necessary to make further reductions for the path sets, but this is much simpler than beginning with the original tree and fashioning the dual of it. The final path sets are $A'D'E'G'$ and $B'D'E'G'$. This indicates that in order for the fault not to occur, it is only necessary for the elements of $ADEG$ or $BDEG$ not to function at the same time.

A shortened version of path-sets problems can be presented. It is possible to derive the path sets from the original cut sets.

Example

Given the minimum cut sets A, BC, BD, CGH of a fault tree, find the path sets.

Solution First we recognize that cut sets presented in this manner are the same as saying:

$$F = A + BC + BD + CGH$$

Therefore, to change the OR gates to AND gates, and vice versa, we simply change the mathematical symbols:

$$F' = A'(B' + C')(B' + D')(C' + G' + H')$$

Now combine terms and reduce as necessary:

$$
\begin{aligned}
F' &= A'(B' + C')(B' + D')(C' + G' + H') \\
F' &= (A'B' + A'C')(B' + D')(C' + G' + H') \\
F' &= (A'B' + A'B'D' + A'B'C' + A'C'D')(C' + G' + H') \\
F' &= (A'B'(1 + D' + C') + A'C'D')(C' + G' + H') \\
F' &= (A'B' + A'C'D')(C' + G' + H') \\
F' &= A'B'C' + A'B'G' + A'B'H' + A'C'D' + A'C'D'G' + A'C'D'H' \\
F' &= A'B'C' + A'B'G' + A'B'H' + A'C'D'(1 + G' + H') \\
F' &= A'B'C' + A'B'G' + A'B'H' + A'C'D'
\end{aligned}
$$

The path sets are thus $A'B'C'$, $A'B'G'$, $A'B'H'$, and $A'C'D'$.

With the preceding examples as guidelines, you should be able to solve the following sample problems.

SAMPLE PROBLEMS

The following fault tree is for Problems 12.1 through 12.10.

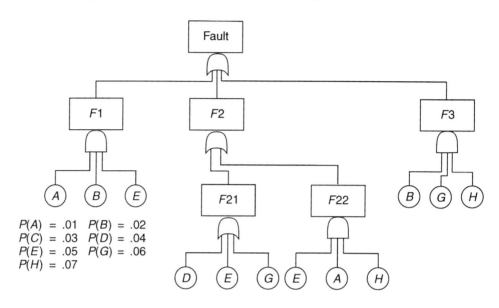

$P(A) = .01$ $P(B) = .02$
$P(C) = .03$ $P(D) = .04$
$P(E) = .05$ $P(G) = .06$
$P(H) = .07$

12.1. Fault = ?

    a. $F1 * F2 * F3$ (223)
    b. $F1 * F2 + F3$ (217)
    c. $F1 + F2 + F3$ (195)
    d. $F1 * F2 + F3$ (213)

12.2. $F1 = ?$

    a. ABE (207)
    b. $A + B + E$ (201)
    c. $F2 + F3$ (276)
    d. $A / B / E$ (270)

12.3. $F2 = ?$

    a. $F21 * F22$ (240)
    b. $DEG + E + A + H$ (264)
    c. $(D + E + G) + (EAH)$ (228)
    d. $F21 + F2$ (224)

12.4. $F3 = ?$

    a. $B * G * H$ (222)
    b. $B + G + H$ (217)
    c. $F1 + F2$ (212)
    d. $F1 * F2$ (211)

12.5. Fault = ?

    a. $F1 * F2 * F3$ (205)
    b. $ABE + (D + E + G + EAH) + BGH$ (201)
    c. $ABE + (DEG + EAH) + BGH$ (199)
    d. $F1 + F2 * F3$ (196)

12.6. Boolean equivalent = ?

    a. $D + E + G + BGH$ (192)
    b. $ABE + D + E + G$ (188)
    c. $D + G$ (185)
    d. $D + E + G$ (180)

12.7. Upper bound on probability of fault = ?

    a. .150219   (176)
    b. .172001   (171)
    c. .150000   (165)
    d. .170000   (254)

12.8. Which of the following equations represents the dual of the fault tree?

    a. $F' = F1' + F2' + F3'$ (250)
    b. $F' = F1'\,F2'(B'G'H')$ (245)
    c. $F' = F1'((D'E'G')(E' + A' + H'))F3'$ (241)
    d. $F = F1F2F3$ (184)

12.9. Which of the following equations is correct for the dual of the fault tree?

    a. $F' = (A' + B' + E')(D'E'G'(E' + A' + H'))(B' + G' + H')$ (234)
    b. $F' = (A'B'E')(D'E'G'(E' + A' + H'))(B'G'H')$ (230)
    c. $F' = (A' + B' + E')(D' + E' + G'(E'A'H'))(B' + G' + H')$ (197)
    d. $F' = (A' + B' + E') + (D'E'G'(E' + A' + H')) + (B' + G' + H')$ (251)

12.10. Which of the following equations represents the path sets of the fault tree?

    a. $B' + D' + G'$ (238)
    b. $D + E + G$ (166)
    c. $D'E'G'$ (175)
    d. $D' + E' + G'$ (181)

12.11. What is the Boolean equivalent of the following equation?

$$F = BD + C + EJH + BCI + EJ + GIL + D + BGHIL + HLM + N$$

    a. $A + C + D + N + EJ + HLM$ (186)
    b. $C + D + N + EJ + GIL + HLM$ (189)
    c. $C + D + N + EJ + GIL + HLM + BGHIL$ (194)
    d. $C + D + EJ + GIL + HLM$ (172)

12.12. What is the Boolean equivalent of the following equation?

$$F = AB + GP + CDE + A + JKN + CDEP + BC + HIP + BCR + BJKLN + B$$

a. $A + B + GP + CDE + JKN + HIP$ (167)
b. $A + B + GP + CDE + BCR + HIP$ (231)
c. $A + B + GP + BC + JKN + HIP$ (253)
d. $A + B + CDE + JKN + HIP$ (248)

12.13. What is the Boolean equivalent of the following equation?

$$F = AZ + AE + BC + ADZ + ADE + BCE + BCEZ + \\ ABDEH + HG + BEZ + ABDE + CEZ$$

a. $AE + BC + HG + BEZ + ECZ$ (244)
b. $AZ + AE + BC + HG + ECZ$ (240)
c. $AZ + AE + BC + HG + BCE + CEZ$ (237)
d. $AZ + AE + BC + HG + BEZ + CEZ$ (233)

12.14. Given the cut sets $BE, DE, DJ, EGJ, GHIJ$, what are the path sets?

a. $F' = E'J' + B'D'G' + B'D'J' + D'E'H' + D'E'I'$ (247)
b. $F' = E'J' + B'D'G' + B'D'J' + D'E'G' + D'E'H' + D'E'I'$ (252)
c. $F' = E'J' + B'D' + I'J' + B'D'J' + D'E'G' + D'E'H' + D'E'I'$ (255)
d. $F' = E'J' + B'D'G' + B'D'J' + D'E'G' + D'E'H' + D'E'K'$ (177)

13

Gaming Theory: Maximin Method

Gaming theory encompasses many methods. Here we focus on one—the maximin method. (Different textbooks will refer to the several strategies of this method by various names. Regardless of the name, the result is the same.) To use any strategy in the maximin method, at least one matrix must be generated. The first matrix generated is often called the maximin matrix.

The maximin matrix is not difficult in theory, but obtaining valid data to generate it can be difficult. The matrix consists of three parts:

1. List of choices or alternatives
2. Events likely to occur
3. Results of the alternatives based on the events

First, the left column represents the alternatives from which we can choose. For example, assume management has allowed you to spend $1 million on safety programs. You might conduct research and determine that you can spend the $1 million in five ways. Your first alternative is to spend $500,000 on earthquake safety, $200,000 on fire safety, and $300,000 on tornado safety. Your second alternative is to spend $250,000 on earthquake safety, $200,000 on fire safety, $200,000 on tornado safety, $200,000 on hurricane safety, and $150,000 on product safety. You would determine all of the possible alternatives. You should have at least three alternatives and probably would not want to consider more than five or six. To construct the matrix, letter the alternatives in the left column. With five alternatives, Table 13-1 shows the first column of the matrix.

The second component involves determining which events are likely to occur. Let's say there is a probability of a minor earthquake, a small fire, and a tornado.

TABLE 13-1.
MAXIMIN ALTERNATIVES

A
B
C
D
E

The second set of events that might occur is a major earthquake, no fire, a tornado, and a small hurricane. Continue generating the various mixes of possible realities. Again, don't overburden yourself with more than five or six possibilities. For our example, assume we can identify four possible outcomes. These outcomes are numbered across the top of the matrix. See Table 13-2.

Determining alternatives may not be too difficult, although in reality there are an infinite number of ways in which you could use the money. Determining the four sets of events to consider as true events is much more difficult, but can be based on thorough risk analyses.

The third component of the matrix is even more difficult to generate, but it is the most important part—namely the results of the alternatives based on each specific event. For example, what happens if you choose alternative B and event 3 really occurs?

These expected results are the "meat" of the matrix. They can be expressed in actual dollar amounts or in terms of utility. For example, if you choose alternative A and event 1 occurs, you may determine that you would save the company $2 million in losses, whereas if you choose alternative A and event 2 occurs you might lose $1 million. If you choose to use actual dollar amounts, then 2 million would be entered in the matrix for $A1$ and negative 1 million would be entered in $A2$. If you prefer relative utilities, you might enter a 1 in $A1$ and a -2 in $A2$, indicating that $A1$ saves you as much as was spent, and $A2$ causes you to lose as

TABLE 13-2. MAXIMIN VARIABLES

		EVENTS			
		1	2	3	4
A L T E R N A T I V E	A				
	B				
	C				
	D				
	E				

TABLE 13-3. MAXIMIN MATRIX

		EVENTS			
		1	2	3	4
A L T E R N A T I V E	A	2	−1	3	−2
	B	0	1	2	−2
	C	−3	3	1	2
	D	1	2	−1	0
	E	1	1	0	2

much. Regardless of the method, you must determine a value for every blank in the matrix, as for example, in Table 13-3.

After the matrix is generated, we determine which alternative to select. The first strategy is a conservative one called maximin. In maximin we first circle the lowest number in each alternative (row). We then select the alternative with the highest circled number. See Table 13-4.

Alternative E is selected because 0 is the highest circled number. The logic of this strategy is simple. It represents the conservative or pessimistic view—no matter which alternative is selected, the worst event in that alternative will occur. Therefore, we want to maximize the solution. By selecting alternative *E* we have guaranteed that we will at least break even. If we had selected any other alternative and the worst event under that alternative occurred, we would have lost between 1 and 3 (million dollars or utility). In some matrices, all alternatives may lead to a loss, but the solution will still lead to the maximum of the minimums.

TABLE 13-4. MAXIMIN DECISION

		EVENTS			
		1	2	3	4
A L T E R N A T I V E	A	2	−1	3	(−2)
	B	0	1	2	(−2)
	C	(−3)	3	1	2
	D	1	2	(−1)	0
	→ E	1	1	(0)	2

TABLE 13-5. EXPECTED VALUE PROBABILITIES

		EVENTS			
		1	2	3	4
		P = .3	.1	.4	.2
A L T E R N A T I V E	A	2	−1	3	−2
	B	0	1	2	−2
	C	−3	3	1	2
	D	1	2	−1	0
	E	1	1	0	2

The second strategy is that we have faith in our matrix and want to select the alternative with the best expected value. This strategy is called Expected Value. To use it, we must add to the matrix the probability of each event. See Table 13-5.

Table 13-5 says that there is a 30% probability that event 1 will occur, 10% that event 2 will occur, and so forth. Note that the sum of the probabilities must equal 1 because we are considering that these are all the events that can possibly happen. To determine the values for expected value, multiply each value in the matrix by its respective probability and sum the values in each row. See Table 13-6.

For this strategy we want to optimize the expected value. Therefore, we select alternative A, expecting to gain 1.3 (dollars or utility).

TABLE 13-6. EXPECTED VALUE (EV) DECISION

		EVENTS					
		1	2	3	4		
		P = .3	.1	.4	.2		EV
A L T E R N A T I V E	→A	2	−1	3	−2	2(.3) + −1(.1) + 3(.4) + −2(.2)	= 1.3
	B	0	1	2	−2	0(.3) + 1(.1) + 2(.4) + −2(.2)	= .5
	C	−3	3	1	2	−3(.3) + 3(.1) + 1(.4) + 2(.2)	= .2
	D	1	2	−1	0	1(.3) + 2(.1) + −1(.4) + 0(.2)	= .1
	E	1	1	0	2	1(.3) + 1(.1) + 0(.4) + 2(.2)	= .8

TABLE 13-7. MINIMAX WORKING MATRIX

		EVENTS			
		1	**2**	**3**	**4**
A					
L					
T					
E	*A*	2–2	3–(–1)	3–3	2–(–2)
R	*B*	2–0	3–1	3–2	2–(–2)
N	*C*	2–(–3)	3–3	3–1	2–2
A					
T	*D*	2–1	3–2	3–(–1)	2–0
I					
V	*E*	2–1	3–1	3–0	2–2
E					

The third strategy is called minimax, or expected regret. For this strategy a new matrix, based on the original matrix, must be generated. First we identify the maximum value under each event (column), 2, 3, 3, 2, respectively. Then we subtract each value for that event (in the column) from the maximum number for that event. For example, see Table 13-7.

Table 13-8 represents the new matrix for the minimax strategy. If the matrix is correct, it has no negative numbers and each column has at least one 0. The matrix represents the regret that we would have if we select an alternative and a specific event occurs. For example, if we select alternative *D* and event 3 occurs, we should be mad at ourselves to a degree of 4, because if alternative *A* had been chosen, we would have gained 3 rather than losing the 1 in alternative *D* (see

TABLE 13-8. MINIMAX MATRIX

		EVENTS			
		1	**2**	**3**	**4**
A					
L					
T					
E	*A*	0	4	0	4
R	*B*	2	2	1	4
N					
A	*C*	5	0	2	0
T	*D*	1	1	4	2
I					
V	*E*	1	2	3	0
E					

TABLE 13-9. MINIMAX DECISION

		EVENTS 1	2	3	4
A L T E R N A T I V E					
	A	0	(4)	0	(4)
	B	2	2	1	(4)
	C	(5)	0	2	0
	D	1	1	(4)	2
→	E	1	2	(3)	0

Table 13-8). In minimax, circle the highest number for each alternative (row) and select the alternative that has the lowest circled number.* (See Table 13-9.)

For this example, that means alternative E is selected. This number means we want to minimize the degree of personal regret. No matter what event occurs, we will never regret it more than a degree of 3. Maximin is a conservative strategy, and minimax is the strategy for the Type A personality who always has to achieve maximum results.

Additional strategies could be discussed, but these are the most common.

The question we therefore need to answer is, which alternative should we select? In this example, alternative E was the alternative of choice under two strategies, but this is not always the case. It is possible that the three strategies would give three different selections. It is up to us to determine which strategy we prefer and defend our choice to management.

In this example we can easily justify alternative E because two of the three strategies resulted in that choice. However, if we really believe our data, then expected value is always easily justified. Also, if we select alternative A, we stand to gain a possible 3 (highest value in alternative A) versus a maximum gain of 2 under alternative E. However, the company may be financially strapped and unable to afford a loss. Therefore, the possibility of experiencing either a −1 or −2 loss might be totally unacceptable, and we cannot afford the gamble of alternative A.

The strategies under the maximin method are not perfect. They do not allow us to make perfect decisions, but they do allow us to evaluate different alternatives based on our personal philosophy.

---

*It does not matter if there are two equally low numbers for an alternative. You can circle either or both of them.

SAMPLE PROBLEMS

For the given maximin matrix, determine the correct choice for each of the following strategies.

| | | EVENTS | | | |
		1	2	3	4
		P = .2	.3	.1	.4
A L T E R N A T I V E	A	3	−2	4	−3
	B	0	2	1	−1
	C	−2	2	0	1
	D	0	3	−2	1
	E	−2	2	1	−1

13.1 Under a maximin strategy, which series of numbers (*A* through *E*) are the correct ones to circle?

a. 4 2 2 1 2   (248)
b. −3 −1 −2 0 −2   (252)
c. −2 −1 −2 0 −1   (231)
d. −3 −1 −2 −2 −2   (169)
e. 4 2 2 3 3   (173)

13.2. Which strategy would you select under maximin?

a. *A*   (178)
b. *B*   (183)
c. *C*   (185)
d. *D*   (192)
e. *E*   (258)

13.3. Which series of numbers represents the respective expected values (*A* through *E*)?

a. −.8 .3 .6 1.1 .1   (255)
b. −.8 1.1 .6 1.1 −.1   (253)
c. −.8 .3 1.4 1.1 −.1   (250)
d. −.8 .3 .6 1.1 −.1   (246)
e. −.8 .3 .6 .7 .1   (242)

13.4. Which strategy would you select under expected value?

    a. *A*   (239)
    b. *B*   (236)
    c. *C*   (232)
    d. *D*   (200)
    e. *E*   (196)

13.5. Which series of values (events 1 through 4) would you use to generate the minimax table?

    a. 3 3 4 1   (193)
    b. −2 −2 −2 −3   (187)
    c. 0 2 0 1   (185)
    d. 3 −2 4 −3   (183)
    e. −2 2 1 −1   (201)

13.6. Which strategy would you select under minimax?

    a. *A*   (204)
    b. *B*   (208)
    c. *C*   (212)
    d. *D*   (216)
    e. *E*   (222)

13.7. Considering all of the strategies, which alternative would you select?

    a. *A*   (225)
    b. *B*   (263)
    c. *C*   (268)
    d. *D*   (277)
    e. *E*   (218)

14

Critical Path Method

Assume your company's president walks in and says that management agrees with your concept for a new systems safety program. However, they need to know by next week exactly how long it will take to complete the program. What is your reaction—total panic, a quick guess, or a detailed plan providing an excellent timetable? Hopefully, it is the last. By using the critical path method (CPM), you will be able to provide a good estimate for the schedule requirements of a program.

CPM was first developed for the chemical industry in 1957. While it was being developed, the United States Navy was developing Program Evaluation and Review Technique (PERT). The methods are very similar, and the terms PERT and CPM are often used interchangeably. We discuss CPM and note the significant differences between it and PERT.

Upon completion, CPM provides a wealth of useful information, including the critical path, the earliest and latest starts of each activity, the earliest and latest finishes of each activity, and the amount of slack each activity contains. Each of these aspects will be developed in more detail as we progress.

The critical path model is based on the activities required to complete a program and the relationship of each activity to the other activities. An activity is any piece of a program that can be identified separately and that requires an expenditure of time to complete. For example, assume Table 14-1 represents activities that need to be completed. (These requirements are representative of the concept phase of a systems safety program and are not intended to represent an entire systems safety program.)

Each activity should be identified by a letter and a brief verbal description.

TABLE 14-1. SYSTEMS SAFETY PROGRAM ACTIVITIES

Activity
a. Develop systems safety program plan
b. Develop initial hazard identification
c. Develop preliminary hazard analysis
d. Develop preliminary energy hazard analysis
e. Develop preliminary risk analysis
f. Assign appropriate personnel
g. Conduct audit
h. Develop initial safety design
j. Conduct concept design review

Some programs require numerous activities. If all 26 letters have been used, use double or triple letters, or make alphanumeric combinations.

Next we determine the length of time needed for each activity. There is no mandatory requirement for the units used (hours, days, weeks, etc.), but after a unit has been established it must be the only one used. For example, if we are using days as the base unit, but feel it necessary to include an activity of 4 hours, we simply list it as $\frac{1}{2}$ day. The length of time allotted to each activity is important and must be done carefully. This time can be based on past history, on conversations with employees, or other means. Table 14-2 notes the length of time, in days, for each activity of our example.

Next we determine the last activity that must be completed before the next activity can begin. For example, the preliminary hazard analysis cannot begin until the initial hazard identification is complete. Therefore, activity *b* is the immediate predecessor of activity *c*. A table of predecessors is given in Table 14-3.

TABLE 14-2. CPM DURATIONS

Activity	Duration (days)
a. Develop systems safety program plan	24
b. Develop initial hazard identification	7
c. Develop preliminary hazard analysis	22
d. Develop preliminary energy hazard analysis	8
e. Develop preliminary risk analysis	18
f. Assign appropriate personnel	10
g. Conduct audit	11
h. Develop initial safety design	20
j. Conduct concept design review	3

TABLE 14-3. CPM PREDECESSORS

Activity	Duration (days)	Predecessor
a. Develop systems safety program plan	24	—
b. Develop initial hazard identification	7	—
c. Develop preliminary hazard analysis	22	*b*
d. Develop preliminary energy hazard analysis	8	*b*
e. Develop preliminary risk analysis	18	*d*
f. Assign appropriate personnel	10	*a*
g. Conduct audit	11	*f*
h. Develop initial safety design	20	*c,e,f*
j. Conduct concept design review	3	*g,h*

These predecessors mean that the activity cannot begin until that activity is completed. For example, activities *a* and *b* have no predecessors so they can begin immediately. Activity *e* cannot begin until activity *d* is complete. Initial safety design cannot begin until activities *c*, *e*, and *f* are completed.

Before we discuss the model for our example CPM, we discuss some general rules and practices. Models are developed using activities and events. Activities, previously discussed, are noted in the model by arrows showing direction. Figure 14-1 represents activity *k*. Note that the letter and the activity duration are given above the arrow.

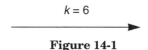

$k = 6$

Figure 14-1

Events are the beginning, or end, points of activities and as such are points in time. They have no duration. For example, although the concept design review itself takes three days, the conclusion of it or the event signifying the signing of the review does not take any real time. Events are numeric. Figure 14-2 shows the beginning and end of activity *m*.

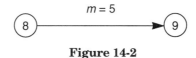

$m = 5$

Figure 14-2

All CPM models begin and end with an event. The first event is the start of the program, and the last event is the completion of the program.

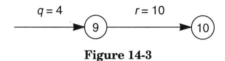

Figure 14-3

Events and activities can be depicted in basically four ways. Figure 14-3 represents the first arrangement. In the figure, r depends on q, or r cannot begin until its predecessor, q, is complete.

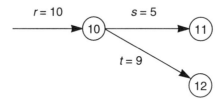

Figure 14-4

Figure 14-4 is a representation that neither s nor t can begin until activity r is complete, but s and t can be conducted simultaneously.

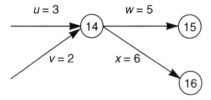

Figure 14-5

Figure 14-5 shows that neither w nor x can begin until both u and v are complete. But u and v, as well as w and x, can run concurrently to each other.

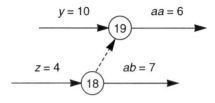

Figure 14-6

The last configuration is denoted by Figure 14-6. Note that ab depends only on z, but that event 19 has two arrows coming into it. The dashed arrow is called a dummy activity. Dummy activities have no duration and merely demonstrate a dependency. Activity aa cannot begin until both y and z are complete. Dummy activities must be considered when determining paths because they mandate the dependency of one activity on another.

To develop the model, start with event 1, start of program. From that event draw an arrow for each activity that does not have a predecessor. Each arrow ends in another event. Then progress from one event to the next, based on which activities are the immediate predecessors of the next activity. It may sound difficult, but it really isn't. However, you may have to redraw some activities initially just to help the flow of the model.

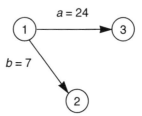

Figure 14-7

Returning to our safety program, let's develop the model. Since neither a nor b has any predecessors, they can both begin immediately (Figure 14-7). Therefore, they are the first arrows in the model, and they emanate from event 1. It does not matter which arrow is on top. In fact, your model may look different from the one developed here, but it is of no consequence. The model will provide the same results.

Since f is the only activity dependent on the completion of activity a, the arrow from f stems from the event terminating activity a. Both c and d have b as their immediate predecessor, so their respective arrows stem from event 2. Activity e can be drawn from event 4 and both c and e conclude in the same event (Figure 14-8). The actual number of an event is basically irrelevant. The convention used here is to proceed from left to right and top to bottom.

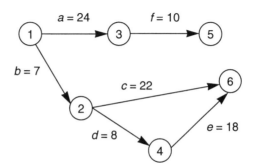

Figure 14-8

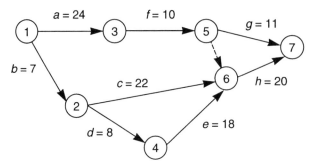

Figure 14-9

Activity *g* depends only on the completion of activity *f*. Activity *h* is a little more difficult. It has *c*, *e*, and *f* as predecessors. If the arrows for activities *c* and *e* were drawn to event 5, the model would indicate that activity *g* depended on *c* and *e*, which it does not. This is a place where the dummy activity is needed to indicate that *g* only depends on *f*, but that *h* depends on *c*, *e*, and *f* (Figure 14-9).

The last activity is *j*. Since it requires both *g* and *h* to be completed, the arrow for *j* stems from the common event for *g* and *h*.

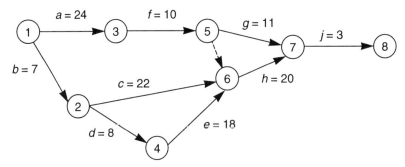

Figure 14-10

Now that the model is fully developed (Figure 14-10), the first major piece of information can be determined. We already determined how long each activity takes, but we really have no idea of the length of the total program. Do we sum all of the times? This is not logical because some activities can overlap. To determine program length, we must determine all paths through the model, by following the arrows through each of the ways that they are connected. Paths can only go in the direction of the arrow. In our example the paths are

     afgj   *bchj*
     afhj   *bdehj*

Although there are many ways to determine the paths, we recommend a systematic way of working from left to right and top to bottom, to avoid skipping a path. Remember a dummy activity is treated just as any other activity.

Add the lengths of the activities along each path to determine the length of time it takes to complete that path. The path lengths are

afgj	48
afhj	57
bchj	52
bdehj	56

By definition, the critical path is that path in which every activity must be completed in the time scheduled in order for the program to remain on schedule. In other words, any delay in an activity on the critical path causes a delay in program completion. Therefore, the critical path is the path that takes the longest time. This seems contrary to our desire for the shortest time for program completion, but remember every path has to be completed before the program can be completed. Therefore, the program cannot be completed until every activity on the longest path is completed. The critical path of our example is *afhj*, and the program will take 57 days to complete.

Now we are ready to determine the wealth of management information that CPM provides. The first calculation is to determine earliest finish (EF) for each activity—the earliest time in which each activity can be completed. Figure 14-11 is identical to Figure 14-10, except that boxes have been added. This convention is standard.

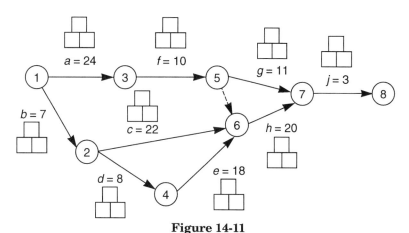

Figure 14-11

The lower left box is where EF is placed. To determine EF begin working from left to right through the model. Event 1 always occurs on day 0. Therefore, the earliest finish for activities that stem from event 1 is computed by adding each activity's duration to zero and placing this number in the appropriate box. To determine EF for any activity stemming from any event, add the activity's duration to the earliest time for which the preceding event can actually occur.

This time takes a little bit of thought. If two or more activities feed into one event, what is the earliest time for which the event can actually occur? It cannot occur until the longest activity is completed. For example, if the earliest finishes of three activities that end in the same event are 4, 8, and 6 respectively, the earliest time that the event can really occur (EE) is 8. Therefore, the computation of earliest finish for any activity stemming from that event is computed by adding the activity's duration to 8.

Since event 6 is the first event with multiple activities leading into it, let's use it as an example. Event 6 must take into consideration the dummy activity. Therefore, event 6 can occur at 29 (*bc* path), 33 (*bde* path), or 34 (*af* path). Since all of the paths leading to event 6 must be completed before it can occur, event 6 cannot occur until day 34. Therefore, the EF of activity *h* is $34 + 20 = 54$. Note that the EF for the last activity on the critical path, activity *j*, must equal the time that we computed for the completion of the program.

Table 14-4 lists the computations of the earliest finishes for our example, and Figure 14-12 shows the earliest finish in the conventional notation.

It is now possible to calculate the latest finish (LF) of each activity. The latest finish is the absolute latest time in which the activity can be completed and the program remain on schedule. To compute LF begin with the last event, end of program. This event must occur in the time determined for the critical path. Thus, LF of event 8 is also the same as the earliest finish of the last activity on the critical path. In our example, LF for event 8 is 57. Therefore, every activity that comes into this event must be completed by that time. The latest finish is noted in the lower right box. Then work from right to left. To determine when the preceding event for any activity must occur, subtract the duration of the activity from the latest finish of that activity. If more than one activity feeds into the same event, choose the appropriate LF of that activity. If the choices were 22, 24, and 18, select the lowest one, 18. This might look like a contradiction, but remember if a higher number were selected then, when working from left to right again, all activities would be extended. For example, the latest that event 5 could

TABLE 14-4. EARLIEST FINISH

$$EF_a = EE_1 + Dur_a = 0 \ + 24 = 24$$
$$EF_b = EE_1 + Dur_b = 0 \ + 7 \ = 7$$
$$EF_c = EE_2 + Dur_c = 7 \ + 22 = 29$$
$$EF_d = EE_2 + Dur_d = 7 \ + 8 \ = 15$$
$$EF_e = EE_4 + Dur_e = 15 + 18 = 33$$
$$EF_f = EE_3 + Dur_f = 24 + 10 = 34$$
$$EF_g = EE_5 + Dur_g = 34 + 11 = 45$$
$$EF_h = EE_6 + Dur_h = 34 + 20 = 54$$
$$EF_j = EE_7 + Dur_j = 54 + 3 \ = 57$$

EE_1–EE_7 equal the earliest that the respective event can actually occur.

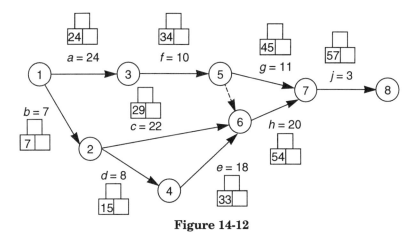

Figure 14-12

occur is either 43 (*jg* path) or 34 (*jh* dummy path). It must occur no later than 34 in order to allow all events to the right of it to occur on time. If event 5 failed to occur before day 43, then the earliest that activity *h* could be completed would be 63 and the schedule would already be overrun. Therefore, activity *f* must have a latest finish time of 34. Figure 14-13 notes the latest finishes.

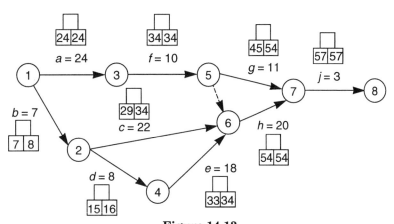

Figure 14-13

One check that will not guarantee accuracy if true but which will guarantee error if not true is that the earliest finish and latest finish for each activity on the critical path must be equal. Since by definition of critical path those activities must be completed on schedule, you should understand why this is true.

We must still compute slack. Slack (S) is simply the amount of time that an activity can slip (overrun) and not affect the schedule. Slack is the difference between the earliest finish and latest finish of each activity and is noted in the top box of the group. Figure 14-14 notes the slack in our safety program.

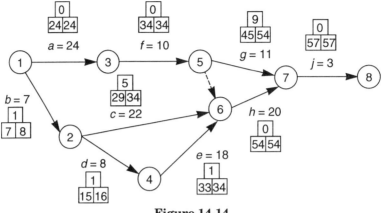

Figure 14-14

Note that slack along the critical path is zero. This means that these activities cannot be delayed at all or the program will not be completed on schedule. The 1 in activity *d* means that this activity can overrun by one day and the program will still be on schedule.

Now the actual model is complete, but we have not developed all of the useful information available. To do so, we reconstruct the model as a table. Table 14-5 is an incomplete table. ES represents the earliest start and is computed by subtracting the duration from the EF for that activity ($ES_a = EF_a - Dur_a$). The new table is Table 14-6. LS represents the latest start and is computed by subtracting the duration from the LF for that activity ($LS_a = LF_a - Dur_a$). Table 14-7 is the completed table.

A simple check for accuracy of ES and LS is that S = LS – ES. If not, there is a mistake somewhere. (It could be in EF or LF, but you know there is a problem somewhere.)

Now we examine the usefulness of this information. Let's start with ES. The earliest start allows us to plan for the allocation of resources. For example, assume we need special equipment that must be purchased in order to complete

TABLE 14-5. INCOMPLETE CPM TABLE

Activity	Duration	ES	EF	LS	LF	S
a.	24		24		24	0
b.	7		7		8	1
c.	22		29		34	5
d.	8		15		16	1
e.	18		33		34	1
f.	10		34		34	0
g.	11		45		54	9
h.	20		54		54	0
j.	3		57		57	0

TABLE 14-6. EARLIEST FINISH

Activity	Duration	ES	EF	LS	LF	S
a.	24	0	24		24	0
b.	7	0	7		8	1
c.	22	7	29		34	5
d.	8	7	15		16	1
e.	18	15	33		34	1
f.	10	24	34		34	0
g.	11	34	45		54	9
h.	20	34	54		54	0
j.	3	54	57		57	0

TABLE 14-7. COMPLETED CPM TABLE

Activity	Duration	ES	EF	LS	LF	S
a.	24	0	24	0	24	0
b.	7	0	7	1	8	1
c.	22	7	29	12	34	5
d.	8	7	15	8	16	1
e.	18	15	33	16	34	1
f.	10	24	34	24	34	0
g.	11	34	45	43	54	9
h.	20	34	54	34	54	0
j.	3	54	57	54	57	0

activity *g*, and we know that it takes 10 days from time of purchase to delivery. Then we need not place the order right away because the equipment cannot be used until day 34. We could wait until around day 20 to 24 to order this equipment. Appropriate planning of earliest start could alleviate cash flow problems or contribute to the wise use of other resources and storage space.

The earliest finish allows us to look forward to the reallocation of resources that may be needed for more than one activity. It tells us the earliest date that we can expect to release these resources from a specific activity.

The latest start is useful in several ways. Assume we have been notified that certain resources for an activity have been delayed. If we are not rapidly approaching the latest start, then the overall schedule may be unaffected. Also, if an activity on the critical path is delayed, we might be able to transfer resources from another activity that does not have to start yet in order to bring the critical activity back on schedule. Latest start can also be used in the allocation of resources in times of need. For example, activity *g* could be delayed from starting until day 43 if the company is experiencing problems with resources.

The latest finish is much the same as the latest start. If a critical activity is behind schedule and resources are being expended on a noncritical activity whose latest finish is some time away, resources may be borrowed from the non-critical activity in order to bring the critical activity back on schedule.

Slack is the overall amount of time we can delay each activity and not delay the schedule. This means that if critical activity h is two days behind schedule, we can borrow two days worth of resources from activity g and not change the schedule.

Caution must be maintained in rescheduling resources because any change in the slack of one activity may influence the slack on later activities. For example, if activity b is delayed one day and uses its slack, then activity c can no longer really start on day 7 and therefore has only four days of slack remaining.

Now that the entire model can be seen, it should be obvious how useful CPM can be. It is not perfect and it depends on the veracity of the information originally generated for the prediction of the duration and flow of activities.

Because the duration of each activity was generated with only a "best guess," CPM cannot be used to make statistical predictions about the accuracy of various dates within the model. Such predictions require a PERT model. It is beyond the scope of this guide to discuss PERT in detail, but it is mentioned because some persons use CPM and call it PERT. This is a violation of the true statistical advantages of PERT. The major difference between CPM and PERT is in the technique used to compute the duration of activities. Instead of using one "best guess" as in CPM, PERT generates an optimistic estimate (OE) of duration, a "best guess" (BG), and a pessimistic estimate (PE) of duration. The actual time scheduled for duration then is based on the formula

$$\frac{OE + (4 * BG) + PE}{6}$$

After the durations are established, CPM and PERT are very similar. However, the variance incorporated in PERT durations allows statistical analysis to be conducted on the model.

SAMPLE PROBLEMS

14.1. For the given table for a safety program, which of the following critical path networks most accurately represents this data? (Duration is in days.)

Activity	Duration	Predecessor
a	12	—
b	3	—
c	5	b
d	2	b
e	2	a
f	10	a,c
g	7	d
h	5	e,f,g

a. (188)

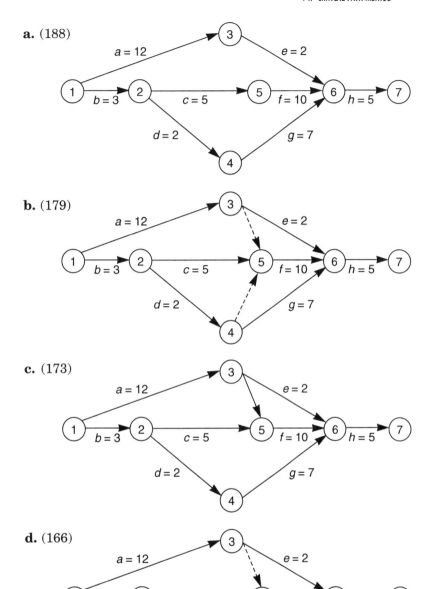

b. (179)

c. (173)

d. (166)

14.2. Using the correct solution from Problem 14.1, which is the correct list of paths through the system?

a. *aefh   afh   bcfh   bdgh*   (277)
b. *aeh   bcfg   bdfh   bdgh*   (270)
c. *aeh   afh   bcfh   bdgh*   (233)
d. *aeh   afh   bceh   bdgh*   (263)

14.3. Which is the critical path?

a. *bcfh*   (267)
b. *afh*   (246)
c. *aeh*   (219)
d. *bceh*   (194)

14.4. Which set of earliest finishes (EF) is correct?

a. $EF_a = 12$ $EF_b = 3$ $EF_c = 8$ $EF_d = 6$ (211)
b. $EF_c = 8$ $EF_d = 5$ $EF_e = 12$ $EF_f = 20$ (204)
c. $EF_e = 14$ $EF_f = 22$ $EF_g = 12$ $EF_h = 27$ (232)
d. $EF_a = 3$ $EF_b = 4$ $EF_c = 8$ $EF_d = 10$ (222)

14.5. Which set of latest finishes (LF) is correct?

a. $LF_a = 13$ $LF_f = 23$ $LF_e = 23$ $LF_h = 28$ (227)
b. $LF_a = 12$ $LF_c = 12$ $LF_f = 22$ $LF_h = 27$ (262)
c. $LF_a = 11$ $LF_d = 14$ $LF_f = 21$ $LF_h = 26$ (243)
d. $LF_a = 12$ $LF_b = 13$ $LF_e = 22$ $LF_h = 27$ (277)

14.6. Using the correct solutions from 14.3 and 14.4, which slack times are correct?

a. $S_a = 0$ $S_c = 4$ $S_f = 0$ $S_h = 0$ (260)
b. $S_a = 0$ $S_b = 4$ $S_e = 7$ $S_h = 0$ (223)
c. $S_b = 4$ $S_c = 4$ $S_d = 10$ $S_f = 1$ (271)
d. $S_d = 10$ $S_e = 8$ $S_g = 11$ $S_h = 0$ (217)

14.7. Which set of earliest starts (ES) is correct?

a. $ES_a = 0$ $ES_b = 0$ $ES_c = 3$ $ES_d = 4$ (192)
b. $ES_c = 3$ $ES_d = 3$ $ES_e = 13$ $ES_f = 12$ (184)
c. $ES_e = 12$ $ES_f = 12$ $ES_g = 4$ $ES_h = 22$ (205)
d. $ES_a = 0$ $ES_c = 3$ $ES_f = 12$ $ES_g = 5$ (170)

14.8. Which set of latest starts (LS) is correct?

a. $LS_a = 0$ $LS_f = 12$ $LS_e = 20$ $LS_h = 22$ (221)
b. $LS_a = 0$ $LS_c = 8$ $LS_f = 12$ $LS_h = 22$ (259)
c. $LS_b = 3$ $LS_d = 13$ $LS_f = 12$ $LS_h = 21$ (227)
d. $LS_b = 4$ $LS_e = 13$ $LS_g = 12$ $LS_h = 27$ (264)

14.9. What is your best course of action if you suddenly found yourself two days behind schedule in the middle of activity f?

    a. Borrow from h (268)
    b. Transfer resources from e (276)
    c. Borrow from b (279)
    d. Transfer resources from g (278)

15

Practice Problems

The following problems are examples of binomial, multinomial, hypergeometric, and critical path method problems you might encounter. Determine the correct answer for each problem.

15.1. A trucking company experienced the following number of accidents over the last 5 years:

Year	Trips	Acc
1992	5175	6
1991	5125	4
1990	5000	4
1989	4875	6
1988	4825	5

If the company expects to make 5200 trips this year, what is the probability of exactly 5 accidents?

    a. 6.6809 E-17   (287)
    b. 1.6810 E-1   (301)
    c. 1.7487 E-1   (320)
    d. 6.5730 E-17   (311)

15.2. Assume the following is a reliability system. What is the probability of the system's success?

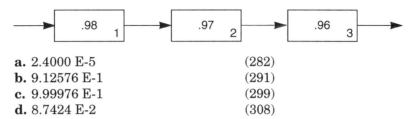

    a. 2.4000 E-5   (282)
    b. 9.12576 E-1   (291)
    c. 9.99976 E-1   (299)
    d. 8.7424 E-2   (308)

15.3. Industry data indicates that a paper mill can expect one fatal accident per 5000 man-years. If a paper mill expects to expend 1000 man-years next year, what is the probability of their experiencing any fatalities?

    a. 1.6378 E-1   (295)
    b. 8.1871 E-1   (296)
    c. 1.8127 E-1   (312)
    d. 1.8128 E-1   (305)

15.4. Your company has experienced a safety problem due to fires caused by faulty resistors. It is known that a parts bin contains 10% faulty ones. If a specific part requires 5 resistors and the parts bin contains 600 resistors, what is the probability that a part will contain no bad resistors?

    a. .950990   (309)
    b. .5893917   (314)
    c. .9508292   (321)
    d. .3298834   (292)

15.5. A company produces a machine that requires 6 identical fuses. In the parts bin, you know there are 2% totally defective fuses and 4% partially defective ones. What is the probability that a machine made from this bin will contain 2 totally defective fuses and 1 partially defective fuse?

    a. 7.6651 E-4   (285)
    b. 5.7600 E-3   (294)
    c. 7.9736 E-4   (298)
    d. 8.4935 E-4   (308)

15.6. Assume the following is an accident system. What is the probability of an accident?

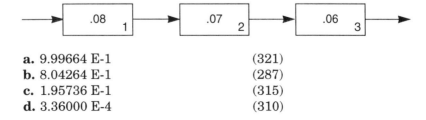

    a. 9.99664 E-1   (321)
    b. 8.04264 E-1   (287)
    c. 1.95736 E-1   (315)
    d. 3.36000 E-4   (310)

15.7. A fire alarm has caused a safety problem due to electrical shorts caused by faulty relays. It is believed that a parts bin contains 20 faulty relays. If a specific part requires 4 relays and the parts bin contains 500 relays, what is the probability that an alarm will contain exactly 1 bad relay?

a. 2.1267 E-4   (326)
b. 8.9358 E-4   (317)
c. 8.4892 E-1   (312)
d. 1.4238 E-1   (304)

15.8. A 4-engine aircraft must have 2 working engines to remain in flight. If the probability of engine failure is .003, what is the probability of an aircraft crashing on any flight due solely to engine failure?

a. 1.0776 E-7   (318)
b. 5.3676 E-5   (302)
c. 1.0768 E-7   (290)
d. 1.1892 E-2   (304)

15.9. Assume the following is a reliability system. What is the probability of failure of the system?

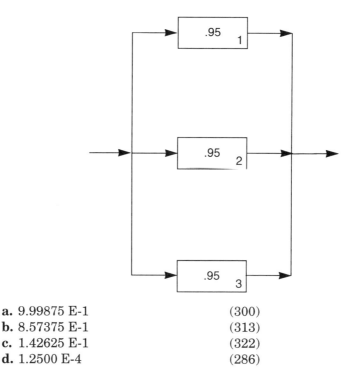

a. 9.99875 E-1   (300)
b. 8.57375 E-1   (313)
c. 1.42625 E-1   (322)
d. 1.2500 E-4   (286)

15.10. A spacecraft has three auxiliary power units (APU). Two are needed for system function. The failure probability of these APUs is .004 per flight. If the spacecraft makes 10 flights per year, what is the probability of system failure due to the APUs?

    a. 4.7862 E-5   (307)
    b. 4.7851 E-4   (298)
    c. 4.7872 E-5   (311)
    d. 4.7808 E-5   (293)

15.11. The probabilities of certain types of accidents in a specific industry are as follows:

Type	Prob
Fatality	.0001
Major	.0009
Minor	.002
None	.9970

What is the probability that in the next year a typical company in the industry will experience 1 fatality, 2 major accidents, 4 minor accidents, and 993 safe trips?

    a. 1.3664 E-24   (319)
    b. 1.3382 E-2   (291)
    c. 1.4687 E-24   (309)
    d. 2.7098 E-4   (327)

15.12. A nuclear reactor control panel contains 8 identical switches. The panel will malfunction if more than 2 switches are faulty. If the panels are made from parts in a bin containing 1000 switches and it is believed that 1.5% of the switches are defective, what is the probability of building a panel that will malfunction due to faulty switches?

    a. 1.4431 E-4   (301)
    b. 1.7524 E-4   (295)
    c. 1.4654 E-4   (284)
    d. 1.4600 E-4   (324)

15.13. Assume the following is an accident system. What is the probability of no accidents?

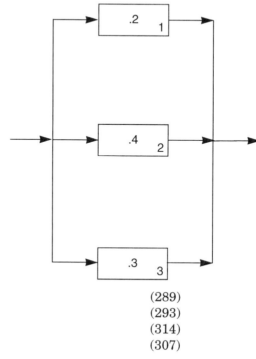

a. .3360	(289)
b. .6640	(293)
c. .02400	(314)
d. .9760	(307)

15.14. Previous experience shows that a particular task has an injury rate of .025 per exposure. If a company expects to have 200 exposures to that task, what is the probability that the company will have less than 6 injuries?

a. .3840300	(283)
b. .6159700	(316)
c. .1480840	(325)
d. .7640540	(299)

15.15. Zervia's Air Force has experienced 10 fatalities, 40 major accidents, and 100 minor accidents in 1000 flights over 4 years. Next year they expect to make 250 flights. What is the probability of their experiencing 2 fatalities, 10 major accidents, and 20 minor accidents?

a. 1.6626 E-3	(318)
b. 1.6626 E-2	(306)
c. 1.6626 E-4	(294)
d. 1.9198 E43	(284)

15.16. A complex pressure release valve contains 5 identical parts. A parts bin contains 150 parts of which 6 are defective. What is the probability of building a valve with any defective parts?

    a. .1869363   (321)
    b. .1742279   (288)
    c. .1846273   (297)
    d. .1698693   (303)

15.17. A smoke detector contains 4 identical resistors. A large bin from which the resistors are selected is known to contain 2% faulty resistors. What is the probability that a smoke detector will contain exactly 1 faulty resistor?

    a. .0775758   (312)
    b. .0755173   (296)
    c. .0023050   (323)
    d. .0752954   (310)

15.18. Assume the following is a reliability system. What is the probability of its success?

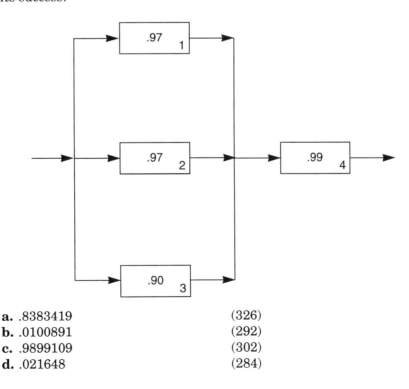

    a. .8383419   (326)
    b. .0100891   (292)
    c. .9899109   (302)
    d. .021648   (284)

15.19. If a company has traditionally experienced .0001 fatalities, .002 major accidents, and .008 minor accidents, what is the probability in the next 500 trips that it will have any accidents?

a. 1.2803 E-3 (306)
b. 9.99498 E-12 (318)
c. 6.2469 E-3 (295)
d. .9937531 (286)

15.20. Assume the following is an accident system. What is the probability of an accident?

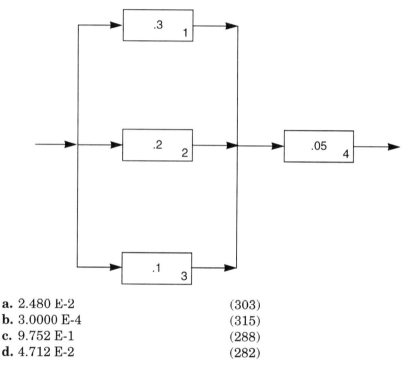

a. 2.480 E-2 (303)
b. 3.0000 E-4 (315)
c. 9.752 E-1 (288)
d. 4.712 E-2 (282)

15.21. An electronic brake contains 4 identical resistors. The probability of a totally defective resistor in the parts bin is .001, and the probability of a partially defective resistor is .015. What is the probability that a brake made with resistors from this bin contains any defective resistors?

a. .0624803 (323)
b. 6.2480 E-2 (310)
c. .0625720 (298)
d. .9375197 (291)

15.22. An electronic component used in a fire suppression system contains 5 identical capacitors. If 2 capacitors fail, the component fails. A parts bin containing .1% faulty capacitors is used in making the component. What is the probability that a component will fail because faulty capacitors were installed?

    a. .999990020   (311)
    b. 9.980 E-6   (294)
    c. .99998   (322)
    d. 2.000 E-5   (304)

15.23. A critical part in a fire suppression system contains 4 identical capacitors. If fewer than 3 capacitors work, the part will fail. Find the probability that the fire suppression system works if made from a bin containing 180 good capacitors and 10 bad ones.

    a. .9999996008   (309)
    b. .80409   (315)
    c. .999589371   (327)
    d. .99999958421   (289)

15.24. Assume the following is a reliability system. What is the probability of success of the system?

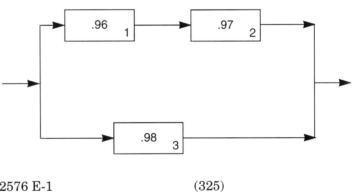

    a. 9.12576 E-1   (325)
    b. 9.98624 E-1   (320)
    c. 1.376 E-3   (317)
    d. 2.4128 E-3   (312)

15.25. Assume the following is an accident system. What is the probability of an accident?

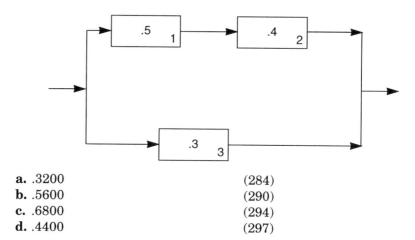

a. .3200   (284)
b. .5600   (290)
c. .6800   (294)
d. .4400   (297)

Appendix A

SOLUTIONS

11.5b. Incorrect. This is just the addition of the two probabilities. This answer does not account for the intersection of the two, which has been counted twice.

Go back to page 115 and select another answer.

12.7c. Correct. Once you find the Boolean equivalent (12.6), simply substitute the individual probabilities. This yields

$$F = D + E + G = .04 + .05 + .06 = .1500$$

Go to the next problem on page 129.

1.9b. Correct. The problem is the notation for the formula for combinations.

$$\binom{100}{10} = \frac{100!}{10!(100-10)!}$$
$$= \frac{100 * 99 * 98 * 97 * 96 * 95 * 94 * 93 * 92 * 91 * 90!}{10 * 9 * 8 * 7 * 6 * 5 * 4 * 3 * 2 * 1 * 90!}$$
$$= 1.7310310 \text{ E}13$$

Go to the next problem on page 7.

4.2a. Incorrect. This solution uses 1's in the r's for the accident categories. This is not what the problem asks.

$$P(1,1,1,47) = \frac{50!}{1! \, 1! \, 1! \, 47!}(.001)^1(.014)^1(.025)^1(.96)^{47}$$
$$= \frac{50 * 49 * 48 * 47!}{1 * 1 * 1 * 47!}(.001)(.014)(.025)(1.46807402\text{E-}1)$$
$$= 117,600 * 5.13825907 \text{ E-}8 = .00604259266$$

Go back to page 35 and select another answer.

2.1c. Incorrect. You probably know what you are doing, but you made an arithmetic error.

Go back to page 17 and select another answer.

9.5a. Incorrect. This is the probability of both I and III failing at the same time. This is close to being correct. What must happen now?

$$P(f)_I * P(f)_{III} = .28 * .454 = .12712$$

Go back to page 90 and select another answer.

11.2c. Incorrect. This is the value of the intersection. This is needed to answer the question. Think about how you will use it.

Go back to page 115 and select another answer.

14.1d. Correct. This network properly shows the relationships among the various activities. It is not necessary for your diagram to look exactly like this, but since this was a multiple choice question it is the only correct solution offered. You properly placed the dummy to indicate the dependency of f on the completions of both a and c. You are ready to analyze the network.

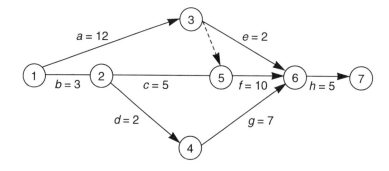

Go to the next problem on page 151.

12.10b. Incorrect. This might help you get started, but it is nowhere close to being correct.

Go back to page 129 and select another answer.

9.8d. **Correct.** The easiest way to solve for success of a parallel system, regardless of whether it is a reliability system or accident system, is to determine the one way that the system will fail and subtract that from 1. For Part I to fail, all three blocks must fail.

$$P(s) = 1 - P(f)_1 * P(f)_2 * P(f)_3 = 1 - .95 * .96 * .97$$
$$= 1 - .88464 = .11536$$

The network can now be redrawn as

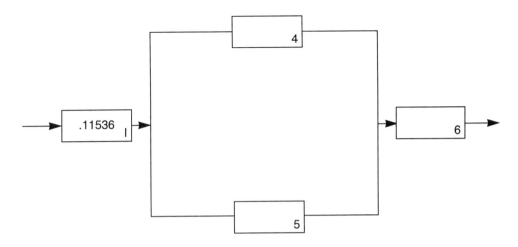

Go to the next problem on page 91.

12.12a. **Correct.** One way to derive this answer

$$F = \cancel{AB} + \cancel{GP} + \cancel{CDE} + A + \cancel{JKN} + \cancel{CDEP} + \cancel{BC} + \cancel{HIP} + \cancel{BCR} + \cancel{BJKLN} + B$$
$$F = A(B+1) + GP + CDE + JKN + CDEP + BC + HIP + BCR + BJKLN + B$$
$$F = A + B\ (C\!+\!CR\!+\!JKLN\!+\!1) + GP + CDE + JKN + CDEP + HIP$$
$$F = A + B + GP + CDE\ (1\!+\!P) + JKN + HIP$$
$$F = A + B + GP + CDE + JKN + HIP$$

Go to the next problem on page 130.

8.10c. **Correct.** You found the correct value for $\chi^2_{.01;6}$ and divided by $2T$. This answer means that based on the tested rate of .003333, you are 98% confident that the true error rate is less than .01401. You recognized that error must be divided by 2 on a two-tailed test.

$$1 - CI \quad = \alpha = 1 - .98 = .02$$
$$\chi^2_{.01;6} \quad = 16.812$$
$$\frac{16.812}{2 * 600} = \frac{16.812}{1200} = .01401$$

Go to the next problem on page 71.

1.9c. **Incorrect.** This is almost correct. An arithmetic error was done in the manual computation. You forgot to multiply by 91.

Go back to page 7 and select another answer.

11.17c. **Correct.** After drawing the diagram and filling in the lines, you can see that the only place where B intersects D and they intersect F' is the intersection of B and D. This is computed by multiplying the two probabilities since they are independent.

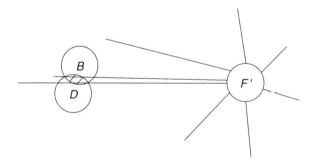

$$P(B \cap D) \cap P(F') = P(B) * P(D)$$
$$.25 * .05 = .0125$$

Go to the next problem on page 117.

9.7c. **Correct.** This is the probability of failure for block 3, that is, the probability of the block functioning so as to prevent an accident.

$$P(f)_3 = 1 - P(s)_3 = 1 - .03 = .97$$

Go to the next problem on page 91.

1.5a. **Incorrect.** Review the rules for rounding.

Go back to page 6 and select another answer.

3.2b. **Correct.** This is the probability of no accidents.

$$P(0) = \binom{1000}{0}(.002)^0(.998)^{1000} = \frac{1000!}{0!1000!} * 1 * (1.35064522 \text{ E} - 1)$$
$$= 1 * 1 * (1.35064522 \text{ E} - 1) = .135064522 = 1.35064522 \text{ E} - 1$$

Go to the next problem on page 31.

10.8c. **Incorrect.** At a minimum, this list contains a set that is not a cut-set (4,5,6) and is missing at least one minimum cut-set.

Go back to page 100 and select another answer.

13.1d. **Correct.** You circle the lowest number in each row. This is the worst thing that can happen under that alternative.

Go to the next problem on page 137.

3.5a. **Incorrect.** This is a dependent problem. This solution is for an independent problem.

$$P(1) = \binom{8}{1}(.02)^1(.98)^7 = \frac{8!}{1!7!}(.02)(.868125533)$$
$$= \frac{8 * 7!}{1 * 7!}(.0173625107)$$
$$= .138900085$$

Go back to page 31 and select another answer.

11.6a. Correct. From the diagram you can see that lines are everywhere. Since you want the union, you want everything with a line in it, so you have the entire universe, which is 1.

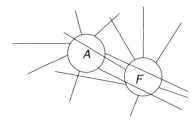

Go to the next problem on page 116.

10.3d. Incorrect. The failure of 4,8 will not even cut the system.

Go back to page 100 and select another answer.

14.7d. Correct. ES is simply EF – Dur. The table you should generate at this point appears as

Activity	Dur.	ES	EF	LS	LF	S
a.	12	0	12		12	0
b.	3	0	3		7	4
c.	5	3	8		12	4
d.	2	3	5		15	10
e.	2	12	14		22	8
f.	10	12	22		22	0
g.	7	5	12		22	10
h.	5	22	27		27	0

Go to the next problem on page 153.

7.6c. **Correct.** First draw a picture to determine what is desired. The next step is to determine the z in order to solve the equation. Since you want a 40% probability, you need the area between 2200 and 2100 to equal .10000. Since x is to the right of $\bar{x}$, you know z is positive. The nearest value without going over in the table for .10000 is .09871. (You could interpolate, but it is not necessary for this problem.) This is the z_{rep} for .25, which is then your z.

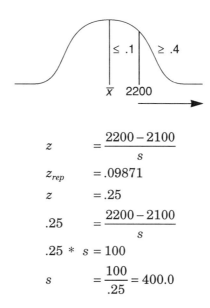

$$z = \frac{2200 - 2100}{s}$$

$$z_{rep} = .09871$$

$$z = .25$$

$$.25 = \frac{2200 - 2100}{s}$$

$$.25 * s = 100$$

$$s = \frac{100}{.25} = 400.0$$

This completes the normal sample problems.

12.7b. **Incorrect.** This number is made up. Quit guessing. What do you need to do before substituting numbers into the equation?

Go back to page 129 and select another answer.

1.9d. **Incorrect.** This is the correct answer for a permutation (order matters). The notation is for a combination. Be careful not to push the permutation button instead of the combination button on the calculator.

Go back to page 7 and select another answer.

5.5d. **Correct.** You appropriately determined that this problem is not dependent. Although it appears to be dependent since it is selecting alarms out of a bin, there is no total for the alarms in the bin. Without the total you cannot work the problem with a dependent formula. Therefore, you make one of two assumptions—there are so many that it does not matter, or someone is replacing them as rapidly as you take one out. You also correctly noted that it is the multinomial, since there must be three categories.

$$P(1,2,5) = \frac{8!}{1!\,2!\,5!}(.012)^1(.028)^2(.96)^5$$

$$= \frac{8 * 7 * 6 * 5!}{1 * 2 * 1 * 5!} * .012 * .000784 * .8153726976$$

$$= \frac{336}{2} * 7.67102634\,\text{E-}6 = 168 * 7.67102634\,\text{E-}6$$

$$= .00128873242$$

This completes the hypergeometric sample problems.

11.14d. Incorrect. This answer is made up. Draw the diagram. What intersects where, and what is the ultimate question?

Go back to page 117 and select another answer.

12.11d. Incorrect. At a minimum, this solution has omitted a term for which there is no way of reducing it. Be careful in your factoring.

Go back to page 129 and select another answer.

8.10b. Incorrect. This answer fails to add 2 to the $2F$ for the degrees of freedom on an upper bound.

Go back to page 71 and select another answer.

6.5a. Correct. Draw a picture to help you determine what to do. Since none of the historic data from Problem 6.4 has changed, and t is the same ($t = 2 * 300 = 600$), the only problem is determining r.

$$P(0) + P(1) + P(2) + \ldots + P(\infty) = 1$$

$$P(< 1) \qquad = P(0) + P(1)$$

$$P(0) \qquad = e^{-t/m} = e^{-600/900} = e^{-.6666666} = .5134171$$

$$P(1) \qquad = \frac{(t/m)^r e^{-t/m}}{r!} = \frac{(600/900)^1 e^{-.6666666}}{1!}$$

$$= .6666666 * .5134171 = .3422781$$

$$P(0) + P(1) = .8556952$$

Go to the next problem on page 49.

8.2a.   Incorrect. You may be thinking properly, but made an arithmetic error to obtain this answer.

$$1 - CI = \alpha = 1 - .9 = .1$$

Go back to page 70 and select another answer.

14.1c.   Incorrect. This diagram is almost correct. However, what is the name of the activity between events 3 and 5? Every activity must be labeled. This is an improper drawing of the necessary connection between events 3 and 5, but we wouldn't call anyone a dummy.

Go back to page 151 and select another answer.

9.15c.   Incorrect. This answer is made up. By now you should be able to do this one.

Go back to page 93 and select another answer.

2.1a.   Incorrect. This answer is the correct answer for the probability of a good resistor.

$$P(\text{good}) = 60 \ (\text{good res})/100 \ (\text{total res}) = .6000$$

Go back to page 17 and select another answer.

1.6d.   Incorrect. This is the correct answer for 6! Be more careful.

Go back to page 7 and select another answer.

11.2d.   Incorrect. This is the probability of the union of B and C. If you are guessing, draw the picture. If not, be more careful in selecting the values to insert.

Go back to page 115 and select another answer.

13.1e.   Incorrect. Some of these numbers do not even appear in the respective row. Think about what you need to do to use the maximin strategy.

Go back to page 137 and select another answer.

1.10a. **Incorrect.** This answer is simply 4 times 20, which is not the formula for a permutation.

Go back to page 7 and select another answer.

11.17b. **Incorrect.** It appears that you may know what you are doing, but you made an arithmetic error. Be a little more careful.

Go back to page 117 and select another answer.

8.11c. **Incorrect.** This answer is derived by making so many mistakes that you probably did not compute it but are guessing. Use the formulas.

Go back to page 71 and select another answer.

6.2a. **Incorrect.** This answer can be derived using the wrong t ($t = 1000$). Check what t is in this problem.

Go back to page 48 and select another answer.

7.1c. **Correct.** After drawing a picture to determine what is desired, the first thing is to determine the z value. Since x is to the left of $\bar{x}$ and you want the probability of $f < x$, subtract z_{rep} from .5.

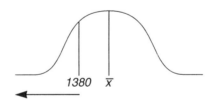

$$z \quad = \frac{1380 - 1400}{50} = \frac{20}{50} = .4$$

$$z_{rep} \qquad = .15541$$

$$P(f < 1380) = .5 - .15541 = .34459$$

Go to the next problem on page 61.

1.3c. **Correct.** You determined the appropriate power and remembered that numbers greater than 5 are rounded up.

Go to the next problem on page 6.

9.6c. Incorrect. This is close, but examine the power.

Go back to page 90 and select another answer.

2.2a. Incorrect. This is the probability of getting exactly one bad resistor in one of either two ways—a good then a bad or a bad then a good. Either way the probability is

$$P(\text{good}) * P(\text{bad}) = .6 * .4 = .2400$$
or
$$P(\text{bad}) * P(\text{good}) = .4 * .6 = .2400$$

Go back to page 17 and select another answer.

12.10c. Correct. Once you drew the dual, you developed the equations and reduced to the Boolean equivalent. One way of doing this is

$$
\begin{aligned}
F' &= F1'F2'F3' \\
F1' &= A' + B' + E' \\
F2' &= F21'F22' \\
F3' &= B' + G' + H' \\
F21' &= D'E'G' \\
F22' &= E' + A' + H' \\
F' &= (A' + B' + E')(D'E'G'\,(E' + A' + H'))(B' + G' + H')
\end{aligned}
$$

Multiplying the middle term yields

$$
\begin{aligned}
F' &= (A' + B' + E')(D'E'G' + A'D'E'G' + D'E'G'H')(B' + G' + H') \\
F' &= (A' + B' + E')(D'E'G'\,(1 + A' + H'))(B' + G' + H') \\
F' &= (A' + B' + E')(D'E'G')(B' + G' + H') \\
F' &= (A'D'E'G' + B'D'E'G' + D'E'G')(B' + G' + H') \\
F' &= (D'E'G'(A' + B' + 1))(B' + G' + H') \\
F' &= D'E'G'(B' + G' + H') \\
F' &= B'D'E'G' + D'E'G' + D'E'G'H' \\
F' &= D'E'G'(B' + 1 + H') \\
F' &= D'E'G'
\end{aligned}
$$

If you had developed a Boolean equivalent of the original tree and were confident of the solution, you could have used the dual of the Boolean equivalent tree to develop your initial equations for the path sets. The dual of the Boolean equivalent of the original tree is

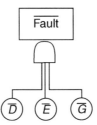

Go to the next problem on page 129.

9.8c. Incorrect. You might have the right idea, but you found the wrong thing to subtract from 1. This just answers the probability of everything except blocks 1, 2, and 3 working at the same time. That is not what is needed to answer the probability of success for a parallel system.

$$1 - P(s)_1 * P(s)_2 * P(s)_3 \ = \ 1 - .05 * .04 * .03 = 1 - .00006$$
$$= .99994$$

Go back to page 91 and select another answer.

4.2b. Incorrect. Quit guessing. This answer is made up.

Go back to page 35 and sclcct another answer.

12.7a. Incorrect. This answer is computed by performing all of the mathematical functions indicated by the original fault tree. You cannot begin substituting numbers into the formula until you have reduced the equation to its Boolean equivalent.

$$F \ = \ ABE + D + E + G + EAH + BGH$$
$$F \ = \ .01(.02)(.05)+.04+.05+.06+.05(.01)(.07)+.02(.06)(.07) = .150219$$

Go back to page 129 and select another answer.

6.4d. **Incorrect.** This answer was obtained by inverting t and m (using m/t instead of the correct t/m).

$$P(0) \neq e^{-m/t} = e^{-900/600} = e^{-1.5} = .22313016$$
$$1 - P(0) = 1 - .22313036 = .77686984$$

Go back to page 49 and select another answer.

1.2a. **Incorrect.** When the decimal point is moved to the right, the power of 10 becomes negative.

Go back to page 6 and select another answer.

12.14d. **Incorrect.** To obtain this answer you have to be guessing because there is a path set here with a letter that does not appear anywhere in the original equation. At least try to work the problems before selecting an answer.

Go back to page 130 and select another answer.

1.10d. **Correct.** It could be done on a calculator using the permutation function. If done manually,

$$\frac{n!}{(n-x)!} = \frac{20!}{(20-4)!} = \frac{20 * 19 * 18 * 17 * 16!}{16!} = 116,280$$

Go to the next problem on page 7.

11.5d. **Incorrect.** This is the probability of the intersection of the two and is needed to obtain the correct answer. What must you do with this probability?

Go back to page 115 and select another answer.

7.6b. **Incorrect.** This is the correct answer if z equals .26. That is close, but think about what that z does to the 40% guarantee.

$$.26 = \frac{2200 - 2100}{s}$$
$$.26 * s = 100$$
$$s = 384.6$$

Go back to page 62 and select another answer.

5.5c. Incorrect. This answer is made up. What kind of problem is this—dependent or independent? Be careful on your answer to that question.

Go back to page 40 and select another answer.

13.2a. Incorrect. This is the worst solution under maximin. After circling the lowest number in each row, which one should you select?

Go back to page 137 and select another answer.

10.3c. Incorrect. The failure of 3,7 alone would cut the system, so it is not necessary for 5 to fail.

Go back to page 100 and select another answer.

8.10a. Incorrect. This is the error rate found in the test. This is what you want to develop a confidence interval around.

$$\text{tested } \lambda = 2/600 = .0033333$$

Go back to page 71 and select another answer.

2.14c. Incorrect. This is the probability of 2 bad resistors. You need to perform a couple more steps.

$$P(2) = \frac{\binom{10}{2}\binom{990}{4}}{\binom{1000}{6}} = \frac{45 * 39{,}782{,}707{,}965}{1.368173 \text{ E}15}$$
$$= 1.30847595 \text{ E-}3$$

Go back to page 22 and select another answer.

11.3a. Incorrect. This answer would be correct if the groups intersected, but they don't. Draw the picture. After filling in the individual groups, you want where the lines cross.

Go back to page 115 and select another answer.

1.7d. Incorrect. This answer is made up.

Go back to page 7 and select another answer.

11.15a. **Incorrect.** This answer is made up. Draw the diagram and think a little harder.

Go back to page 117 and select another answer.

11.17a. **Incorrect.** This answer is made up. Draw the diagram.

Go back to page 117 and select another answer.

9.8b. **Incorrect.** This is the probability of all three blocks failing, but that is not exactly the question. What do you need to do now?

$$P(f)_1 * P(f)_2 * P(f)_3 = .95 * .96 * .97 = .88464$$

Go back to page 91 and select another answer.

3.2c. **Incorrect.** This is the probability of 1 accident found in Problem 3.1d, page 31.

Go back to page 31 and select another answer.

6.3a. **Incorrect.** This is the correct answer for exactly 1 failure. This does not answer the entire question. Draw the picture for assistance.

$$P(1) = \frac{(t/m)^r e^{-t/m}}{r!} = \frac{(700/800)^1 e^{-.875}}{1!}$$
$$= .875 * .4168620 = .3647543$$

Go back to page 48 and select another answer.

8.2b. **Incorrect.** The error must be divided by 2 on a two-tailed test.

Go back to page 70 and select another answer.

14.1b. **Incorrect.** According to this network, a, c, and d must be complete before f can begin. A dummy shows a dependency on a previous activity's completion. The table does not indicate that f is dependent of d's completion. Examine the table more carefully.

Go back to page 151 and select another answer.

9.5c. **Incorrect.** This is the probability of everything except both Parts I and III working. Try again.

$$1 - P(s)_{\mathrm{I}} * P(s)_{\mathrm{III}} = 1 - .72 * .546 = 1 - .39312 = .60688$$

Go back to page 90 and select another answer.

12.6d. **Correct.** This answer was derived as follows:

$$
\begin{aligned}
F   &= F1 + F2 + F3 \\
F1 &= ABE \\
F2 &= F21 + F22 \\
F3 &= BGH \\
F21 &= D + E + G \\
F22 &= EAH \\
F   &= ABE + F21 + F22 + BGH \\
F   &= ABE + D + E + G + EAH + BGH \\
F   &= D + E(AB + 1 + H) + G(1 + BH) \\
F   &= D + E + G
\end{aligned}
$$

Go to the next problem on page 129.

9.15d. **Correct.** This is the probability of failure of a reliability block when MTBF data is provided. Solve for $P(0)$ and subtract from 1.

$$P(f) = 1 - P(0) = 1 - e^{-t/m} = 1 - e^{-80/4.5\mathrm{E}3}$$
$$1 - e^{\,1.777777\mathrm{E}\text{-}2} = 1 - .982379315$$
$$= 1.76206854\ \mathrm{E}\text{-}2$$

This completes the event system sample problems.

2.2d. **Incorrect.** It appears that you might know what you are doing, but you made an arithmetic mistake.

Go back to page 17 and select another answer.

1.11c. **Incorrect.** This is the solution for a permutation, but the problem specifies that order does not matter. Therefore, it is a combination.

Go back to page 7 and select another answer.

11.5c. Correct. There is obviously an overlap (intersection) of the two groups. Also, from the given information, you realized that the groups are dependent to the degree that $P(E \cap F) = .06$.

$$P(E \cup F) = P(E) + P(F) - P(E \cap F)$$
$$.15 + .2 - .06 = .29$$

Go to the next problem on page 116.

4.2c. Correct. If you are predicting no accidents, that means each of the accident categories must be 0.

$$P(0,0,0,50) = \frac{50!}{0!0!0!50!}(.001)^0(.014)^0(.025)^0(.96)^{50}$$
$$= 1 * 1 * 1 * 1 * 1.29885794\,E\text{-}1$$
$$= .129885794$$

Go to the next problem on page 35.

12.10d. Incorrect. This is close. The only problem is that once the dual is drawn, all gates are reversed and you should not arrive at this exact answer.

Go back to page 129 and select another answer.

10.1a. Incorrect. The failure of 1,6 will not cut the system because it could work through 2,3,4 among other ways.

Go back to page 99 and select another answer.

5.5b. Incorrect. You were correct to assume that it is independent and thus the multinomial was appropriate, but you left out one piece. Examine the question again and perform your checks. You also made a transposition error.

$$P(1,2) = \frac{8!}{2!\,1!}(.012)^2(.028)^1$$
$$= 20,160 * .000144 * .028 = .08128512$$

Go back to page 40 and select another answer.

9.10d. Incorrect. It appears that you are guessing. This is the probability of everything except all of the major blocks failing at the same time.

$$1 - P(f)_{\mathrm{I}} * P(f)_{\mathrm{II}} * P(f)_6 = 1 - .88464 * .8742 * .9$$
$$1 - .6960170592 = .3039829408$$

Go back to page 91 and select another answer.

8.9d. Correct. You found the correct value for $\chi^2_{.05;26}$ and divided by $2T$. This answer means that based on the tested rate of .019090, you are 95% confident that the true error rate is less than .017675.

$$1 - CI = \alpha = 1 - .95 = .05$$
$$2F + 2 = 2 * 12 + 2 = 26$$
$$\chi^2_{.05;26} = 38.885$$
$$\frac{38.885}{2 * 1100} = \frac{38.885}{2200} = .017675$$

Go to the next problem on page 71.

13.2b. **Correct.** Circle the lowest number in each option (row), and then select the highest of those circled. Thus, no matter which alternative actually occurs, you lose no more than 1.

	1	2	3	4
	.2	.3	.1	.4
A	3	−2	4	(−3)
B	0	2	1	(−1)
C	(−2)	2	0	1
D	0	3	(−2)	1
E	(−2)	2	1	−1

Go to the next problem on page 137.

10.3b. **Correct.** The simultaneous failure of 3 and 6 will cut the system, and there is no other smaller cut-set containing both blocks.

Go to the next problem on page 100.

1.12a. **Correct.** Without doing anything, if you remember the rule, in a combinations notation, anything over 0 equals 1.

$$\binom{1000}{0} = \frac{1000!}{0!(1000-0)!} = \frac{1000!}{0!\,1000!} = 1$$

Go to the next problem on page 7.

13.5d. **Incorrect.** This is just the first alternative's numbers. Why would they be better than any other alternative? Think about what you need to generate the new table.

Go back to page 138 and select another answer.

6.2c. **Incorrect.** This answer is made up. What is the formula for Poisson? What is m? What is t?

Go back to page 48 and select another answer.

12.8d. Incorrect. This is close. The problem is that once the dual is drawn, all terms become prime, which means "not."

Go back to page 129 and select another answer.

11.3b. Incorrect. This answer is made up. Draw the picture.

Go back to page 115 and select another answer.

8.5a. Incorrect. Degrees of freedom and F are not synonymous. Check the formula.

Go back to page 70 and select another answer.

9.8a. Incorrect. This is the probability of blocks 1, 2, and 3 working. This is only one way that the blocks could work in order for there to be success for the accident system. Think about the easiest way to solve a parallel network.

$$P(s)_1 * P(s)_2 * P(s)_3 = .05 * .04 * .03 = .00006$$

Go back to page 91 and select another answer.

14.7b. Incorrect. At least ES_e is incorrect. ES is simply EF – Dur. To obtain the correct answer for ES_e, take EF_e and subtract Dur_e from it:

$$ES_e = EF_e - Dur_e = 14 - 2 = 12$$

Go back to page 152 and select another answer.

6.6d. Incorrect. This answer can be obtained by transposing t and m (using m/t rather than the correct t/m):

$$P(3) \neq \frac{(t/m)^r e^{-t/m}}{r!} = \frac{(450/140)^3 e^{-3.214286}}{3!}$$

$$= \frac{(3.214286)^3 * e^{-3.214286}}{3!} = \frac{33.2088192421 * .0401840265}{6}$$

$$= \frac{1.33446407083}{6} = .222410678$$

Go back to page 49 and select another answer.

11.15b. **Incorrect.** This answer is for $P(B \cap C) \cup P(A)$. Note, however, the original problem has B'.

Go back to page 117 and select another answer.

12.6c. **Incorrect.** This answer is close, but there is at least one cut-set missing.

Go back to page 129 and select another answer.

4.2d. **Incorrect.** We have no idea why you think this one cannot be solved.

Go back to page 35 and select another answer.

9.5d. **Correct.** The easiest way to solve a parallel network is to determine the only way that it can fail and subtract that from 1.

$$P(S) = 1 - P(f)_I * P(f)_{III} = 1 - .28 * .454$$
$$= 1 - .12712 = .87288$$

Go to the next problem on page 90.

1.13a. **Incorrect.** This number is made up.

Go back to page 8 and select another answer.

13.2c. **Incorrect.** You can do better than guaranteeing a loss of 2 if the worst alternative for each solution occurs.

Go back to page 137 and select another answer.

7.6a. **Incorrect.** This answer is made up. How do you determine the z value?

Go back to page 62 and select another answer.

13.5c. **Incorrect.** This selection actually makes no sense at all. Think about what you need to generate the new table.

Go back to page 138 and select another answer.

5.5a. **Incorrect.** You were correct to assume that it is independent and thus the multinomial is appropriate, but you left out one piece. Examine the question again and perform your checks.

$$P(1,2) = \frac{8!}{1!\,2!}(.012)^1(.028)^2$$
$$= 20,160 * .012 * .000784 = .18966528$$

Go back to page 40 and select another answer.

12.11a. **Incorrect.** A term in this equation is not even in the original equation. Quit guessing. Carefully reduce the equation using the principles of Boolean logic.

Go back to page 129 and select another answer.

10.3a. **Incorrect.** The failure of 3,7 alone would cut the system so it is not necessary for 1 to fail.

Go back to page 100 and select another answer.

8.9c. **Incorrect.** If you really got this answer, you probably misread the table. Carefully line up the degrees of freedom and error rate. You went down 1 degree of freedom.

Go back to page 71 and select another answer.

6.2d. **Correct.** You first determine the values of λ or m. Since nothing in the historic data has changed from Problem 6.1, they are the same. Then determine that t is 1500 and r is 2.

$$P(2) = \frac{(t/m)^r e^{-t/m}}{r!} = \frac{(1500/800)^2 e^{-1500/800}}{2!}$$
$$= \frac{(1.875)^2 e^{-1.875}}{2!} = \frac{3.515625 e^{-1.875}}{2!}$$
$$= \frac{3.515625 * .1533550}{2!} = \frac{.5391386}{2} = .2695693$$

Go to the next problem on page 48.

3.2d. **Incorrect.** This is the probability of 2 accidents found in Problem 3.1c, page 239.

Go back to page 31 and select another answer.

9.7d. **Incorrect.** This is the probability of failure for block 1. It may help if you fill in the entire table.

Go back to page 91 and select another answer.

8.2c. **Correct.** The error allowed is .1, but this must be divided by 2 since it is a two-tailed test.

Go to the next problem on page 70.

10.1b. **Incorrect.** The failure of 2,3 will not cut the system because it could work through 1,6,7,8.

Go back to page 99 and select another answer.

11.16d. **Incorrect.** This answer is made up. Draw the diagram. Where is the union between E and C? Where is the union between this and A'?

Go back to page 117 and select another answer.

4.3a. **Incorrect.** This answer forgets to divide by the 2! and 4!.

$$P(1,2,4,993) = \frac{1000!}{1! \, 2! \, 4! \, 993!}(.0004)^1(.001)^2(.002)^4(.9966)^{993}$$
$$= .212949498$$

Go back to page 35 and select another answer.

7.1d. Incorrect. This is the correct answer for the area between x and $\bar{x}$, but that is not what you are seeking. Draw the correct diagram and reevaluate what you are seeking.

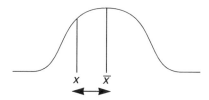

Go back to page 61 and select another answer.

1.13d. Incorrect. There is no / or ÷ in the notation. This notation is for a combination.

Go back to page 8 and select another answer.

14.1a. Incorrect. The table states that activities a and c must be complete before f can begin. This diagram fails to account for this.

Go back to page 151 and select another answer.

12.6b. Incorrect. Look at *ABE*. Can it not be reduced? Are there other mistakes?

Go back to page 129 and select another answer.

13.5b. Incorrect. These are the lowest numbers under each event. What numbers do you need to generate the minimax table? Think about the strategy behind the minimax decision.

Go back to page 138 and select another answer.

11.5a. Incorrect. This answer is almost correct. You probably drew the picture correctly and thought you were doing the mathematics appropriately. However, there is the problem of dependency. Think a little harder.

Go back to page 115 and select another answer.

5.4d. Incorrect. Close, but you made an error in transposition.

$$P(1,2,5) \neq \frac{\binom{6}{2}\binom{14}{1}\binom{480}{5}}{\binom{500}{8}} = \frac{\frac{6!}{2!\,4!}\,\frac{14!}{1!\,13!}\,\frac{480!}{5!\,475!}}{\frac{500!}{8!\,492!}}$$

$$= \frac{\frac{30}{2} * \frac{14}{1} * \frac{2.49534144\ E13}{120}}{\frac{3.69247042\ E21}{40,320}} = \frac{15 * 14 * 207,945,120,096}{9.15791275\ E16}$$

$$= \frac{4.36684752\ E13}{9.15791275\ E16} = 4.76838734\ E\text{-}4$$

Go back to page 40 and select another answer.

9.6a. Correct. This can be determined by subtracting the probability of success found in Problem 9.5d, page 185, from 1.

$$1 - P(s)_{\text{sys}} = 1 - .87288 = .12712$$

Go to the next problem on page 91.

11.3c. Correct. From the drawing, you should be able to see that they do not intersect. Therefore the correct answer is 0.

Go to the next problem on page 115.

7.5d. Incorrect. This answer is made up.

Go back to page 61 and select another answer.

12.11b. Correct. This answer was derived by:

$$F = B\!\!\!\!/D + \cancel{C} + \cancel{EJH} + BCI + EJ + \cancel{GIL} + \cancel{D} + \cancel{BGIHL} + HLM + \cancel{N}$$
$$F = C(1 + BI) + D(B + 1) + N + EJ(H + 1) + GIL(1 + BH) + HLM$$
$$F = C + D + N + EJ + GIL + H\Lambda M$$

Go to the next problem on page 130.

2.1b. **Incorrect.** You made two mistakes. First, you tried to determine the probability of good resistors, not bad ones. You also made an arithmetic error ($60/100 \neq .06$).

Go back to page 17 and select another answer.

11.16c. **Incorrect.** This is the answer for $P(E \cup C) \cup P(A)$. Read the problem more carefully.

Go back to page 117 and select another answer.

8.9b. **Incorrect.** This answer used χ^2 for a two-tailed test. Do not divide the error for a one-tailed test.

$$\chi^2_{.025;26} = 41.923$$

Go back to page 71 and select another answer.

10.2d. **Incorrect.** The simultaneous failure of 4,5,8 would cut the system.

Go back to page 99 and select another answer.

1.12c. **Incorrect.** Quit guessing.

Go back to page 7 and select another answer.

1.3a. **Incorrect.** The power is correct, but you failed to remember the rounding rules.

Go back to page 6 and select another answer.

8.2d. **Incorrect.** This answer would be extremely hard to get without guessing. What is the error allowed, and what must be done to it before using the table?

Go back to page 70 and select another answer.

1.8d. **Correct.** It could be done on a calculator using the factorial button. Manually,

$$3! * 3! = 3 * 2 * 1 * 3 * 2 * 1 = 36$$

Go to the next problem on page 7.

5.4c. Correct. To arrive at this answer you must first decide how many of each type alarm you have. This is computed by multiplying the probabilities by the total of 500. You realize that you are missing a category and logically assume it is for good alarms. The problem is dependent, so use the hypergeometric distribution.

$$P(1,2,5) = \frac{\binom{6}{1}\binom{14}{2}\binom{480}{5}}{\binom{500}{8}} = \frac{\frac{6!}{1!5!}\frac{14!}{2!12!}\frac{480!}{5!475!}}{\frac{500!}{8!492!}}$$

$$= \frac{\frac{6*5!}{1*5!}\frac{14*13*12!}{2*12!}\frac{480*479*478*477*476*475!}{5*4*3*2*1*475!}}{\frac{500*499*498*497*496*495*494*493*492!}{8*7*6*5*4*3*2*1*492!}}$$

$$= \frac{6*\dfrac{182}{2}*\dfrac{2.49534144\ E13}{120}}{\dfrac{3.69247042\ E21}{40,320}} = \frac{6*91*207,945,120,096}{9.15791275\ E16}$$

$$= \frac{1.13538036\ E14}{9.15791275\ E16} = 1.23978071\ E\text{-}3$$

Go to the next problem on page 40.

4.3b. Correct. First determine the individual probabilities on any given exposure. For fatalities, that is 2/5000 = .0004. Since the probabilities do not sum to 1, there must be a missing category—safe exposures. Its probability is $1 - P(\text{sum}_{acc})$. Obviously this accounts for the remaining 993 exposures also.

$$P(1,2,4,993) = \frac{1000!}{1!2!4!993!}(.0004)^1(.001)^2(.002)^4(.9966)^{993}$$

$$= \frac{1000*999*\ldots*993!}{1*2*24*993!}(4\ E\text{-}4)(1\ E\text{-}6)(1.6\ E\text{-}11)(3.39810\ E\text{-}2)$$

$$= \frac{9.79174267\ E20}{48}2.1747865\ E\text{-}22$$

$$= 2.03994639\ E19*2.1747865\ E\text{-}22$$

$$= .00443644787$$

Go to the next problem on page 35.

1.2b. Incorrect. There are two mistakes here. The power should be negative, and it is also off by one place.

Go back to page 6 and select another answer.

14.7a. Incorrect. At least ES_d is incorrect. ES is simply EF – Dur. To obtain the correct answer for ES_d, take EF_d and subtract Dur_d from it:

$$ES_d = EF_d - Dur_d = 5 - 2 = 3$$

Go back to page 152 and select another answer.

10.1c. Incorrect. The failure of 4,5,6 will not cut the system because it could work through 1,3,7,8 or 2,3,7,8.

Go back to page 99 and select another answer.

12.6a. Incorrect. Look at *BGH*. Can it not be reduced? Are there other mistakes?

Go back to page 129 and select another answer.

1.6c. Incorrect. This answer was derived by adding instead of multiplying.

$$5 + 4 + 3 + 2 + 1 = 15 \neq 5!$$

Go back to page 7 and select another answer.

13.2d. Incorrect. You can do better than guaranteeing that you will lose 2 if the worst alternative for each solution occurs.

Go back to page 137 and select another answer.

6.4c. Incorrect. This answer is derived by rounding too early.

$$P(0) = e^{-t/m} = e^{-600/900} = e^{-.667} = .513246009$$
$$1 - P(0) = 1 - .513246009 = .486753991$$

Go back to page 49 and select another answer.

11.15d. **Correct.** This is the probability of the intersection of B' and C in union with A. The picture is most helpful here. Once you determine where B' and C intersect, the easiest way to compute it is to subtract the intersection of B and C from C. Then add this to A.

$$P(B' \cap C) \cup P(A) = P(C) - P(B \cap C) + P(A)$$
$$= .1 - (.25 * .1) + .11 = .185$$

Go to the next problem on page 117.

11.4d. **Incorrect.** This answer is made up. Draw the picture.

Go back to page 115 and select another answer.

9.6b. **Incorrect.** This is close, but there is a rounding error.

Go back to page 90 and select another answer.

13.5a. **Correct.** These are the highest numbers under each event. What do you do with these numbers now?

Go to the next problem on page 138.

2.13b. **Incorrect.** This is the probability of everything except 2 defective hats.

$$1 - P(2) = .995022388$$

Go back to page 22 and select another answer.

4.3c. **Incorrect.** This answer is made up.

Go back to page 35 and select another answer.

11.12b. Incorrect. This answer is made up. Draw the diagram. What is the question?

Go back to page 116 and select another answer.

3.3a. Correct. Remember, the probability of any is always $1 - P(0)$.

$$P(0) + P(1) + P(2) + \ldots + P(1000) = 1$$
$$P(\text{any}) = 1 - P(0)$$
$$P(0) \quad = \binom{1000}{0}(.002)^0(.998)^{1000}$$
$$= \frac{1000!}{0! \; 1000!} * 1 * (1.35064522 \, \mathrm{E} \text{-} 1)$$
$$= 1 * 1 * (1.35064522 \, \mathrm{E} \text{-} 1) = .135064522 = 1.35064522 \, \mathrm{E} \text{-} 1$$
$$1 - P(0) = 1 - .135064522 = .864935478$$

Go to the next problem on page 31.

12.11c. Incorrect. At a minimum, cannot *GIL* be factored from *BGHIL*, thus eliminating the need for *BGHIL*? Remember, the Boolean equivalent represents the same thing as the minimum cut-sets. Since *GIL* is contained in *BGHIL*, *BGHIL* cannot be a minimum cut-set.

Go back to page 129 and select another answer.

2.2b. Correct. The only way of getting no bad resistors is to get good resistors on both draws. Thus, the Multiplication Law is in effect.

$$P(\text{good}) * P(\text{good}) = .6 * .6 = .36$$

Go to the next problem on page 17.

14.3d. Incorrect. This is not even a legitimate path. If you selected this answer, by means other than guessing, stop and review Problem 14.2 before you continue.

Go back to page 152 and select another answer.

10.2c. **Correct.** The failure of 1,5,7 will not cut the system because, even if through no other way, it could work through 2,3,4.

Go to the next problem on page 99.

11.16b. **Incorrect.** This is the correct answer for $E \cup C$.

Go back to page 117 and select another answer.

8.9a. **Incorrect.** This is the error rate of the test itself (12/1100), which is what you are not confident of and want to develop a confidence interval around. Check the formula and find the correct χ^2 for the information provided.

Go back to page 71 and select another answer.

1.7c. **Incorrect.** This answer is 4! and is basically made up.

Go back to page 7 and select another answer.

5.4b. **Incorrect.** You forgot to use your check to see if you have everything. What is missing?

$$P(1,2) = \frac{\binom{6}{1}\binom{14}{2}}{\binom{500}{8}} = \frac{\frac{6!}{1!\,5!}\frac{14!}{2!\,12!}}{\frac{500!}{8!\,492!}}$$

$$= \frac{546}{9.15791275 \ \text{E}16} = 5.9620572 \ \text{E-}15$$

Go back to page 40 and select another answer.

12.1c. **Correct.** The gate is an OR gate, and OR equals +. Add the number of boxes coming into this gate.

Go to the next problem on page 128.

9.7b. **Incorrect.** This is the probability of failure for block 5. It may help if you fill in the entire table.

Go back to page 91 and select another answer.

8.3a. Incorrect. You may have made an arithmetic error. Be careful when converting percentages into decimals.

Go back to page 70 and select another answer.

12.5d. Incorrect. No gate can have both addition and multiplication stemming from it.

Go back to page 129 and select another answer.

11.4c. Incorrect. This is the correct answer for the union of C and D, but the question calls for the intersection.

Go back to page 115 and select another answer.

13.4e. Incorrect. This is one of the worst options under expected value.

Go back to page 138 and select another answer.

3.3b. Incorrect. This number is made up.

Go back to page 31 and select another answer.

2.3c. Correct. This answer is derived by determining that there are two ways of getting exactly one bad resistor: bad, good; or good, bad. Either one is correct, so use the Addition Rule.

$$P(\text{bad}) * P(\text{good}) = .4 * .6 = .2400$$
$$\text{or}$$
$$P(\text{good}) * P(\text{bad}) = .6 * .4 = .2400$$
$$P(1_{\text{bad}}) = .2400 + .2400 = .4800$$

Go to the next problem on page 17.

10.2b. Incorrect. The simultaneous failure of 3,6,7 would cut the system.

Go back to page 99 and select another answer.

9.5b. **Incorrect.** This is the probability of both Parts I and II working. It is only one way a parallel network can work. Think about what is needed.

$$P(s)_\mathrm{I} * P(s)_\mathrm{III} = .72 * .546 = .39312$$

Go back to page 90 and select another answer.

4.3d. **Incorrect.** You probably miscalculated the probabilities per exposure, or you failed to account for the safe exposures.

Go back to page 35 and select another answer.

11.16a. **Correct.** The figure shows lines everywhere except at A. You want the union, so you want everything with a line in it, which is the entire universe minus A, namely $1 - .11$.

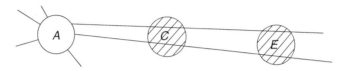

Go to the next problem on page 117.

8.8d. **Incorrect.** This answer is a guess, or you are not reading the formula correctly. Both the error and the degrees of freedom are wrong if you obtained this value from the table. This is $\chi^2_{.1;22}$.

Go back to page 71 and select another answer.

12.9c. **Incorrect.** Look at the terms for $F21'$. What kind of gate should the one below $F21$ have been switched to for the dual?

Go back to page 129 and select another answer.

3.1b. Incorrect. This is the probability of everything except exactly 1 accident.

$$p = 4/2000 = .002$$

$$P(1) = \binom{1000}{1}(.002)^1(.998)^{999} = \frac{1000!}{1!\ 999!}(.002)(1.35335193E\text{-}1)$$

$$= \frac{1000\ *\ 999!}{1\ *\ 999!}(2.70670386\ E\text{-}4) = 1000(2.70670386\ E\text{-}4)$$

$$= .270670386$$

$$1 - P(1) = 1 - .270670386 = .729329614$$

Go back to page 31 and select another answer.

11.4a. Correct. The diagram shows that C and D intersect. There is no indication that they are dependent, so multiply the individual probabilities.

$$P(C \cap D) = P(C) * P(D) = .1 * .05 = .005$$

Go to the next problem on page 115.

6.6a. Incorrect. This answer is made up. How many test hours were there? What is m? With this information you should be able to solve this problem.

Go back to page 49 and select another answer.

9.7a. Incorrect. This is the probability of failure for block 6. It may help if you fill in the entire table.

Go back to page 91 and select another answer.

6.6c. **Correct.** This is the probability of exactly 3 failures. Determine λ or m, and t. There are 12 systems tested, each for 150 hours. Thus, total test time is 1800 hours.

$\lambda = .002\overline{222}$ or $m = 450$ $t = 140$

$$P(0) + P(1) + P(2) + P(3) + \ldots + P(\infty) = 1$$

$$P(3) = \frac{(t/m)^r e^{-t/m}}{r!} = \frac{(140/450)^3 e^{-.31111111}}{3!}$$

$$= \frac{(.31111111)^3 \ast e^{-.31111111}}{3!} = \frac{.0301124829 \ast .732632468}{6}$$

$$= \frac{.0220613826}{6} = .0036768971$$

Go to the next problem on page 49.

3.3c. **Incorrect.** Quit guessing. This number is made up.

Go back to page 31 and select another answer.

12.5c. **Incorrect.** Some of these terms are correct, but examine *DEG* a little more closely.

Go back to page 129 and select another answer.

8.3b. **Incorrect.** This would be correct if it were a two-tailed test, which it is not.

Go back to page 70 and select another answer.

10.2a. **Incorrect.** The simultaneous failure of 1,2,3 would cut the system.

Go back to page 99 and select another answer.

1.11a. **Incorrect.** This answer is simply 27 divided by 3. The problem calls for a combination.

Go back to page 7 and select another answer.

13.4d. **Correct.** Multiply the value for each option by the probability of that event. Then add those numbers. Select the option with the highest value.

	1	2	3	4	
	.2	.3	.1	.4	
A	3	−2	4	−3	$3(.2) + -2(.3) + 4(.1) + -3(.4) = -.8$
B	0	2	1	−1	$0(.2) + 2(.3) + 1(.1) + -1(.4) = .3$
C	−2	2	0	1	$-2(.2) + 2(.3) + 0(.1) + 1(.4) = .6$
D	0	3	−2	1	$0(.2) + 3(.3) + -2(.1) + 1(.4) = 1.1$
E	−2	2	1	−1	$-2(.2) + 2(.3) + 1(.1) + -1(.4) = -.1$

Go to the next problem on page 138.

11.4b. **Incorrect.** This is for the union without considering the intersection. Draw the picture and try again.

Go back to page 115 and select another answer.

9.6d. **Incorrect.** This has several mistakes in it. Think about how to find the failure if you already know success.

Go back to page 90 and select another answer.

2.5b. **Incorrect.** This is the probability of everything except all bad resistors.

$$P(2_{bad}) = .1600$$
$$1 - P(2_{bad}) = 1 - .1600 = .8400$$

Go back to page 17 and select another answer.

11.13a. **Incorrect.** This answer is made up. Draw the diagram. Where is the intersection of A and E? Where is the union between this intersection and C?

Go back to page 117 and select another answer.

2.8a. **Incorrect.** This represents the probability of one of several ways that 2 bad resistors could be drawn. One way is

$$P(b,b,g) = \frac{40}{100} * \frac{39}{99} * \frac{60}{98} = \frac{93,600}{970,200} = .096474954$$

Go back to page 18 and select another answer.

12.2b. **Incorrect.** What kind of gate leads into $F1$? Check what you do with that kind of gate.

Go back to page 128 and select another answer.

9.14a. **Correct.** This is the probability of failure of the block. Remember that when the MTBF is used, $P(f)$ represents the probability of success for an accident system. It will be much easier in these cases to use a table and label it.

$$P(0) = e^{-t/m} = e^{-75/1.4E2} = e^{-5.35714286E-1}$$
$$= .585251104$$

Go to the next problem on page 93.

13.5e. **Incorrect.** This is just the last alternative's numbers. Why would they be better than any other alternative? Think about what you need to generate the new table.

Go back to page 138 and select another answer.

8.8c. **Incorrect.** This is a one-tailed test. What error should you find in the table? This is $\chi^2.1;24$.

Go back to page 71 and select another answer.

12.5b. **Correct.** $F1 = ABE$, $F2 = D + E + G + EAH$, and $F3 = BGH$. Since $F = F1 + F2 + F3$, add the three.

Go to the next problem on page 129.

6.6b. **Incorrect.** This answer is made up. What is t? What is m?

Go back to page 49 and select another answer.

7.5c. Incorrect. This answer could be derived by finding the z_{rep} for .95, but $z \neq .95$. How do you find the z value?

Go back to page 61 and select another answer.

9.1a. Incorrect. This is the probability of failure for block 5. Fill in the table and it might help.

Go back to page 90 and select another answer.

11.8a. Correct. From the diagram you can tell that the only place the lines intersect is within A.

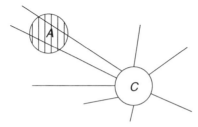

Go to the next problem on page 116.

2.15b. Incorrect. This is the probability of everything except 2 bad resistors.

$$P(2) \quad - \frac{\binom{10}{2}\binom{990}{4}}{\binom{1000}{6}} = \frac{45 * 39{,}782{,}707{,}965}{1.368173 \text{ E}15}$$

$$= 1.30847595 \text{ E-}3$$
$$1 - P(2) = .998691524$$

Go back to page 22 and select another answer.

3.3d. Incorrect. This is close, but there is a rounding error.

Go back to page 31 and select another answer.

11.19c. Incorrect. This answer is made up. Draw the diagram.

Go back to page 117 and select another answer.

2.5d. **Correct.** Remember the easiest way to determine the probability of any is always the probability of $1 - P(0)$.

$$P(0) \ = \ .6 * .6 = .3600$$
$$1 - P(0) \ = \ 1 - .3600 = .6400$$

Go to the next problem on page 18.

2.10a. **Incorrect.** This is the probability of no bad resistors. You need to do one more step. Draw a picture to assist in determining what to do.

$$P(0_{\text{bad}}) = \frac{60}{100} * \frac{59}{99} * \frac{58}{98} = \frac{205,320}{970,200} = .211626469$$

Go back to page 18 and select another answer.

4.4a. **Correct.** At first this appears to be a dependent problem which could not be solved by the multinomial formula. However, there is no indication of the total number of parts in the bin, so no determination of dependency can be made. You assume it is independent and proceed from there. Since you need 9 parts for your component, the number of good parts must be 4 in order for the total parts needed to equal 9.

$$P(2,3,4) = \frac{9!}{2!\,3!\,4!}(.1)^2(.2)^3(.7)^4$$
$$= \frac{9 * 8 * 7 * 6 * 5 * 4!}{2 * 3 * 2 * 4!}(.01)(.008)(.2401)$$
$$= \frac{15,120}{12}\,1.9208\ \text{E-5} = 1260 * 1.9208\ \text{E-5}$$
$$= .02420208$$

This completes the multinomial sample problems.

2.14b. **Incorrect.** This is the probability of 1 bad resistor. You need to perform a couple more steps.

$$P(1) = \frac{\binom{10}{1}\binom{990}{5}}{\binom{1000}{6}} = \frac{10 * 7.845150\text{E}12}{1.368173\ \text{E}15}$$
$$= .0573403239$$

Go back to page 22 and select another answer.

10.5d. Incorrect. All of the sets are correct except for 2,4,5,6, which will not cut the system.

Go back to page 100 and select another answer.

13.6a. Incorrect. There is one better solution under minimax.

Go back to page 138 and select another answer.

2.6a. Correct. This is the probability of getting exactly 3 bad resistors in 3 draws. This problem is dependent, which means that what happens during one trial (draw) affects the next trial (draw). Therefore,

$$P(3_{bad}) = \frac{40}{100} * \frac{39}{99} * \frac{38}{98} = \frac{59,280}{970,200} = .061100804$$

Go to the next problem on page 18.

5.3c. Incorrect. This is needed for the correct answer, but is not quite there. What does this solution answer?

$$P(0,10) = \frac{\binom{16}{0}\binom{64}{10}}{\binom{80}{10}} = \frac{\frac{16!}{0!\,16!}\frac{64!}{10!\,54!}}{\frac{80!}{10!\,70!}}$$

$$= \frac{151,473,214,816}{1.64649211\,\mathrm{E}12} = .0919975346$$

Go back to page 39 and select another answer.

14.4b. Incorrect. At least EF_e is incorrect. To obtain the correct answer, take EF_a and add the duration of e to it:

$$EF_e = EF_a + Dur_e = 12 + 2 = 14$$

Go back to page 152 and select another answer.

9.13d. **Incorrect.** This is the probability of failure of a reliability block when MTBF data is provided. How would you obtain the correct answer to the problem?

$$P(f) = 1 - P(0) = 1 - e^{-t/m} = 1 - e^{-50/1.2E3}$$
$$1 - e^{-4.16666667E\text{-}2} = 1 - .959189457$$
$$= 4.08105429 \text{ E-2}$$

Go back to page 92 and select another answer.

8.8b. **Incorrect.** Check your degrees of freedom. This is $\chi^2.2;22$

Go back to page 71 and select another answer.

6.4b. **Incorrect.** This answer is derived by rounding too early.

$$P(0) = e^{-t/m} = e^{-600/900} = e^{-.67} = .511708578$$
$$1 - P(0) = 1 - .511708578 = .488291422$$

Go back to page 49 and select another answer.

8.3c. **Correct.** Error is equal to 1 minus confidence. Since this is a one-tailed test, that is the error you use in the table.

Go to the next problem on page 70.

9.1b. **Incorrect.** This is the probability of failure for blocks 3 or 4. Fill in the table and it might help.

Go back to page 90 and select another answer.

14.7c. **Incorrect.** At least ES_g is incorrect. ES is simply EF – Dur. To obtain the correct answer for ES_g, take EF_g and subtract Dur_g from it:

$$ES_g = EF_g - Dur_g = 12 - 7 = 5$$

Go back to page 152 and select another answer.

12.5a. **Incorrect.** Which kind of gate is under the fault?

Go back to page 129 and select another answer.

2.6c. Incorrect. This answer is for the probability of exactly 3 good resistors, assuming that the problem is independent.

$$P(3_{good(ind)}) = .6 * .6 * .6 = .2160$$

Go back to page 18 and select another answer.

11.12d. Correct. Remember, once you have solved the intersection of E and F, that is all you have. Therefore, when you want the union with B, you simply add that intersection with B. Don't forget that E and F are dependent.

$$P(E \cap F) \cup P(B) = .06 + .25 = .31$$

Go to the next problem on page 117.

7.5b. Correct. First draw a picture to determine what is desired. The next step is to determine the z in order to solve the equation. Since you want a 95% probability, you need .45000 of the area to be between 1000 and $\bar{x}$. Since .95000 must be to the right of 1000, you know 1000 is to the left of $\bar{x}$. The nearest value without going under in the table for .45000 is .45053. (You could interpolate, but it is not necessary for this problem.) This is the z_{rep} for 1.65, which is then your z. Since x is to the left of $\bar{x}$, you know that z is negative.

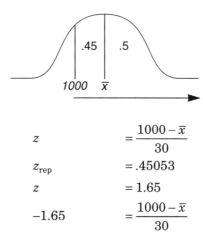

$$z \qquad = \frac{1000 - \bar{x}}{30}$$

$$z_{rep} \qquad = .45053$$

$$z \qquad = 1.65$$

$$-1.65 \qquad = \frac{1000 - \bar{x}}{30}$$

$$-1.65 * 30 \quad = 1000 - \bar{x}$$
$$-49.5 \quad\quad = 1000 - \bar{x}$$
$$-1000 + (-49.5) = -\bar{x}$$
$$1049.5 \quad\quad = \bar{x}$$

Go to the next problem on page 61.

1.8b. **Incorrect.** Remember that you cannot simply multiply factorials.

$$3! * 3! = 3 * 2 * 1 * 3 * 2 * 1$$
$$9! = 9 * 8 * 7 * 6 * 5 * 4 * 3 * 2 * 1$$

Go back to page 7 and select another answer.

12.2a. **Correct.** $F1$ has an AND gate leading into it. Therefore, multiply every-thing that comes into it.

Go to the next problem on page 128.

3.4a. **Incorrect.** This is the probability of exactly 4 accidents.

$$P(4) = \binom{1000}{4}(.002)^4(.998)^{996}$$
$$= \frac{1000!}{4!996!}(1.6\ \text{E-}11)(1.36150463\text{E-}1)$$
$$= \frac{1000 * 999 * 998 * 997 * 996!}{4 * 3 * 2 * 1 * 996!}(2.1784074\ \text{E-}12)$$
$$= \frac{994,010,994,000}{24}(2.1784074\ \text{E-}12)$$
$$= 41,417,124,750(2.1784074\ \text{E-}12) = .0902233713$$

Go back to page 31 and select another answer.

13.6b. Correct. First generate the regret matrix by determining the highest number for each alternative (column) and subtracting each number in that column from the highest value in the column.

	1	2	3	4
	.2	.3	.1	.4
A	3 – 3	3 – –2	4 – 4	1 – –3
B	3 – 0	3 – 2	4 – 1	1 – –1
C	3 – –2	3 – 2	4 – 0	1 – 1
D	3 – 0	3 – 3	4 – –2	1 – 1
E	3 – –2	3 – 2	4 – 1	1 – –1

Now circle the highest value in each option (row) and select the lowest of those circled. This means that no matter which alternative occurs, the most angry you will be at yourself is a factor of 3.

	1	2	3	4
	.2	.3	.1	.4
A	0	(5)	0	4
B	3	1	(3)	2
C	(5)	1	4	0
D	3	0	(6)	0
E	(5)	1	3	2

Go to the next problem on page 138.

11.8b. Incorrect. This answer is made up. Draw the diagram. Where is the intersection?

Go back to page 116 and select another answer.

9.13c. Correct. This is the probability of success for a block in a reliability system where MTBF data is provided. Use the Poisson to determine $P(0)$.

$$P(0) = e^{-t/m} = e^{-50/1.2E3} = e^{-4.16666667E\text{-}2}$$
$$= .959189457$$

Go to the next problem on page 92.

1.9a. Incorrect. There is no / or ÷ in the symbology. This is not a fraction.

Go back to page 7 and select another answer.

8.8a. Correct. You recognized that a one-tailed test does not divide the error, and an upper bound uses $2F + 2$ as the degrees of freedom.

$$\chi^2_{.2;24} = 29.553$$

Go to the next problem on page 71.

2.7a. Incorrect. This probability assumes independency, which is incorrect.

$$P(\text{bad}_{\text{ind}}) * P(\text{good}_{\text{ind}}) * P(\text{bad}_{\text{ind}}) = .4 * .6 * .4 = .09600$$

Go back to page 18 and select another answer.

4.4b. Incorrect. This is close, but it appears that you forgot to divide by 4!.

$$P(2,3,4) \neq \frac{9!}{2!\,3!}(.1)^2(.2)^3(.7)^4$$
$$= .58084992$$

Go back to page 35 and select another answer.

10.6a. Correct. Each of these sets is a minimum cut-set. There are no smaller sets containing all of the blocks in each of these sets.

Go to the next problem on page 100.

2.12d. Incorrect. This is close to the correct answer, but check the rules for rounding.

Go back to page 21 and select another answer.

11.19d. Correct. This answer requires careful thought and arithmetic after drawing the diagram. You must subtract the intersections of B and C, B and D, and C and D. Thus, you subtract the intersection of B, C, and D twice too often so you must add the intersection of B, C, and D back into the equation twice.

$$P(B \cup C) \cup P(D) = P(B) + P(C) + P(D) - P(B \cap C) - P(B \cap D)$$
$$- P(C \cap D) + P(B \cap C \cap D) + P(B \cap C \cap D)$$
$$= .25 + .1 + .05 - (.25 * .1) - (.25 * .05) - (.1 * .05) + (.25 * .1 * .05) + (.25 * .1 * .05)$$
$$= .4 - .025 - .0125 - .005 + .00125 + .00125 = .3600$$

Go to the next problem on page 118.

3.4b. Incorrect. This is the solution for exactly 3 accidents. This is needed, but it is not the final answer.

$$P(3) = \binom{1000}{3}(.002)^3(.998)^{997} = \frac{1000!}{3!997!}(8.0 \text{ E} - 9)(1.35878162 \text{ E} - 1)$$
$$= \frac{1000 * 999 * 998 * 997!}{3 * 2 * 1 * 997!}(1.0870253 \text{ E} - 9)$$
$$= \frac{997,002,000}{6}(1.0870253 \text{ E} - 9)$$
$$= 166,167,000(1.0870253 \text{ E} - 9) = .180627732$$

Go back to page 31 and select another answer.

8.3d. Incorrect. Quit guessing. If you obtained the answer for Problem 8.2, you should not be having trouble with this one.

Go back to page 70 and select another answer.

9.1c. Incorrect. This is the probability of failure for block 2. Fill in the table and it might help.

Go back to page 90 and select another answer.

12.4d. **Incorrect.** Are you guessing? $F1$ and $F2$ are on the same level as $F3$. They have nothing to do directly with $F3$. What kind of gate stems from $F3$?

Go back to page 128 and select another answer.

5.3b. **Incorrect.** This would be correct if the question asked for the probability of any dust or mist cartridges in 9 draws. Reread the question.

$$P(0,0,9) = \frac{\binom{6}{0}\binom{10}{0}\binom{64}{9}}{\binom{80}{9}} = \frac{\frac{6!}{0!6!}\frac{10!}{0!10!}\frac{64!}{9!55!}}{\frac{80!}{9!71!}}$$

$$= \frac{1 * 1 * 27{,}540{,}584{,}512}{231{,}900{,}297{,}200} = .118760454$$

$$= 1 - P(0,0,9) = 1 - .118760454 = .881239546$$

Go back to page 39 and select another answer.

14.4a. **Incorrect.** At least EF_d is incorrect. To obtain the correct answer, take EF_b and add the duration of d to it:

$$EF_d = EF_b + Dur_d = 3 + 2 = 5$$

Go back to page 152 and select another answer.

11.12c. **Incorrect.** This answer is made up. Draw the diagram. Where does E intersect F? Where does this intersection union with B?

Go back to page 116 and select another answer.

9.13b. **Incorrect.** It appears that you are guessing. This answer could be derived by using the wrong t, but how did you make the exact same mistake?

$$P(f) = 1 - P(0) = 1 - e^{-t/m} = 1 - e^{-5/1.2E3}$$

$$= 1 - e^{-4.1666667E-3} = 1 - .995842002$$

$$= 4.15799816\ E{-}3$$

Go back to page 92 and select another answer.

11.8c. **Incorrect.** This is the correct answer for $A' \cup C$. Draw the diagram. Where do the lines intersect (cross)?

Go back to page 116 and select another answer.

13.6c. **Incorrect.** There is one better solution under minimax.

Go back to page 138 and select another answer.

10.6b. **Incorrect.** All of the sets are minimum cut-sets except 5,7,8, which is not even a cut-set.

Go back to page 100 and select another answer.

2.8b. **Incorrect.** This accounts for two of the ways that exactly 2 bad resistors could be drawn, but there are more.

$$P(b,b,g) \quad = \frac{40}{100} * \frac{39}{99} * \frac{60}{98} = \frac{93,600}{970,200} = .096474954$$

$$P(g,b,b) \quad = \frac{60}{100} * \frac{40}{99} * \frac{39}{98} = \frac{93,600}{970,200} = .096474954$$

$$P(b,b,g) + P(g,b,b) \quad = .096474954 + .096474954 = .19295$$

Go back to page 18 and select another answer.

8.7d. **Correct.** It appears that you know how to use the formula and find the correct values in the table.

$$\chi^2_{.05;18} = 28.869$$

Go to the next problem on page 71.

1.10b. **Incorrect.** This is simply 20 divided by 4, which is not the formula for a permutation.

Go back to page 7 and select another answer.

12.4c. **Incorrect.** Are you guessing? $F1$ and $F2$ are on the same level as $F3$. They have nothing to do directly with $F3$. What kind of gate stems from $F3$?

Go back to page 128 and select another answer.

7.5a. Incorrect. Close, but you used an area under the curve that is too low. The area must be greater than the projected area if you are to guarantee a 95% chance of withstanding 1000 psi.

Go back to page 61 and select another answer.

12.1d. Incorrect. No gate can have both addition and multiplication stemming from it.

Go back to page 128 and select another answer.

3.4c. Incorrect. This is the probability of everything except 3 accidents.

$$P(3) = \binom{1000}{3}(.002)^3(.998)^{997}$$

$$= \frac{1000!}{3!\,997!}(8.0\ E\text{-}9)(1.35878162\ E\text{-}1)$$

$$= \frac{1000 * 999 * 998 * 997!}{3 * 2 * 1 * 997!}(1.0870253\ E\text{-}9)$$

$$= \frac{997,002,000}{6}(1.0870253\ E\text{-}9)$$

$$= 166,167,000(1.0870253\ E\text{-}9) = .180627732$$

$$= 1 - P(3) = 1 - .180627732 = .819372268$$

Go back to page 31 and select another answer.

2.4a. Incorrect. This is the probability of getting 2 good resistors.

$$P(\text{good}) * P(\text{good}) = .6 * .6 = .3600$$

Go back to page 17 and select another answer.

5.3a. Correct. You noticed that the number of cartridges needed was changed. This is the correct solution for all organic cartridges. Just as in Problem 5.2, this can be solved as a two-category problem.

$$P(0,0,10) = \frac{\binom{6}{0}\binom{10}{0}\binom{64}{10}}{\binom{80}{10}} = \frac{\frac{6!}{0!6!}\frac{10!}{0!10!}\frac{64!}{10!54!}}{\frac{80!}{10!70!}}$$

$$= \frac{1 * 1 * \frac{64 * 63 * 62 * 61 * 60 * 59 * 58 * 57 * 56 * 55 * 54!}{10 * 9 * 8 * 7 * 6 * 5 * 4 * 3 * 2 * 1 * 54!}}{\frac{80 * 79 * 78 * 77 * 76 * 75 * 74 * 73 * 72 * 71 * 70!}{10 * 9 * 8 * 7 * 6 * 5 * 4 * 3 * 2 * 1 * 70!}}$$

$$= \frac{\frac{5.49666002\,\text{E}17}{3,628,800}}{\frac{5.97479057\,\text{E}18}{3,628,800}} = \frac{151,473,214,816}{1.64649211\,\text{E}12} = .0919975346$$

Then

$$1 - P(0,0,10) = 1 - .0919975346 = .908002465$$

$$P(0,10) \quad = \frac{\binom{16}{0}\binom{64}{10}}{\binom{80}{10}} = \frac{\frac{16!}{0!16!}\frac{64!}{10!54!}}{\frac{80!}{10!70!}}$$

$$= \frac{151,473,214,816}{1.64649211\,\text{E}12} = .0919975346$$

Then

$$1 - P(0,10) = 1 - .0919975346 = .908002465$$

Go to the next problem on page 40.

7.2a. Correct. First sketch the graph to determine what is desired. This is a bit tricky, but if you want the probability of not failing, then you need the area to the right of x because the area to the left represents the probability of failure. Next, determine the z value. Then you can see that since x is to the right of $\bar{x}$ and you want the probability of $f > x$, you must subtract z_{rep} from .5.

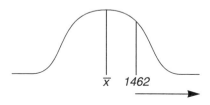

$$\bar{x} \quad 1462$$

$$z \qquad = \frac{1462 - 1400}{50} \quad \frac{62}{50} = 1.24$$

$$z_{rep} \qquad = .39252$$

$$P(f < 1462) = .5 - .39252 = .10748$$

Go to the next problem on page 61.

8.4a. Incorrect. F is used to compute the degrees of freedom, but it is not the degrees of freedom. Check the formula again.

Go back to page 70 and select another answer.

9.1d. Correct. This is the probability of failure for block 1. Fill in the entire table and use it to answer the remaining questions.

$$P(f)_1 = 1 - P(s)_1 = 1 - .9 = .1$$

Go to the next problem on page 90.

2.8c. Incorrect. You probably know what you are doing, but you made a rounding error. (The answer is correct to 4 significant digits, but 5 are given in the answer.)

Go back to page 18 and select another answer.

11.3d. Incorrect. This is the probability of the intersection of E and F. Be more careful.

Go back to page 115 and select another answer.

4.4c. Incorrect. This number is made up.

Go back to page 35 and select another answer.

11.20a. Incorrect. This answer is made up. Draw the diagram and carefully determine what is needed.

Go back to page 118 and select another answer.

13.6d. Incorrect. This is the worst solution under minimax.

Go back to page 138 and select another answer.

9.13a. Incorrect. It appears that you are guessing. This answer could be derived by using the wrong t, but how did you make the exact same mistake?

$$P(0) = e^{-t/m} = e^{-5/1.2E3} = e^{-4.1666667E\text{-}3} = .995842002$$

Go back to page 92 and select another answer.

11.8d. Incorrect. This answer is made up. Draw the diagram. Where is C'? Where is A? Where do they intersect?

Go to the next problem on page 116.

8.7c. Incorrect. On an upper-bound test, make sure you have determined the proper degrees of freedom. This is $\chi^2_{.05;16}$.

Go back to page 71 and select another answer.

1.13c. Correct.

$$\binom{1000}{5} = \frac{1000!}{5!(1000-5)!}$$
$$= \frac{1000 * 999 * 998 * 997 * 996 * 995!}{5 * 4 * 3 * 2 * 1 * 995!}$$
$$= 8.2502912 \text{ E12}$$

This completes the general mathematics sample problems.

6.2b.   Incorrect. This is the correct answer for the probability of exactly 1 failure.

Go back to page 48 and select another answer.

10.6c.   Incorrect. All of the sets are minimum cut-sets except 3,7,8.

Go back to page 100 and select another answer.

12.4b.   Incorrect. What kind of gate stems from $F3$?

Go back to page 128 and select another answer.

14.6d.   Incorrect. At least S_g is incorrect. To compute S subtract EF from LF:

$$S_g = LF_g - EF_g = 22 - 12 = 10$$

Go back to page 152 and select another answer.

12.1b.   Incorrect. Regardless of which kind of gate is under the fault, it is not possible to have addition and multiplication from the same gate.

Go back to page 128 and select another answer.

2.12c.   Incorrect. This is the correct probability for any bad resistors. You performed one too many steps.

Go back to page 21 and select another answer.

5.2d. Correct. In order to determine the probability of all organic cartridges in 9 draws, you want none of the other categories. You can solve it as 3 categories, but doesn't it act as 2—organic vs. not organic? Either way gives the correct solution.

$$P(0,0,9) = \frac{\binom{6}{0}\binom{10}{0}\binom{64}{9}}{\binom{80}{9}} = \frac{\frac{6!}{0!\,6!}\frac{10!}{0!\,10!}\frac{64!}{9!\,55!}}{\frac{80!}{9!\,71!}}$$

$$= \frac{1 * 1 * 27{,}540{,}584{,}512}{231{,}900{,}297{,}200} = .118760454$$

$$P(0,9) = \frac{\binom{16}{0}\binom{64}{9}}{\binom{80}{9}} = \frac{\frac{16!}{0!\,16!}\frac{64!}{9!\,55!}}{\frac{80!}{9!\,71!}}$$

$$= \frac{1 * 27{,}540{,}584{,}512}{231{,}900{,}297{,}200} = .118760454$$

Go to the next problem on page 39.

13.7e. Incorrect. There is no real justification for selecting this alternative. Are you serious or just guessing?

Go back to page 138 and select another answer.

7.2b. Incorrect. This is the probability of $f < 1462$.

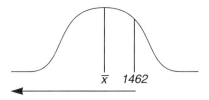

Go back to page 61 and select another answer.

9.12d. Incorrect. It appears that you are guessing. This answer could be derived by using the wrong t, but how did you make the exact same mistake?

$$P(f) = 1 - P(0) = 1 - e^{-t/m} = 1 - e^{-100/2.3E5}$$
$$= 1 - e^{-4.34782609E\text{-}4} = 1 - .999565312$$
$$= 4.34688104 \text{ E-4}$$

Go back to page 92 and select another answer.

2.9c. Incorrect. There are two problems here. You assumed independency. Second, this answer is only part of the correct answer. Draw the picture to help determine the solution.

$$P(b,b,b) = .4 * .4 * .4 = .06400$$

Go back to page 18 and select another answer.

11.12a. Incorrect. This answer is made up. Draw the diagram. Carefully analyze each step.

Go back to page 116 and select another answer.

14.3c. Incorrect. This path only requires 19 days. There is at least one longer path. Remember the critical path is the longest path because the job cannot be done until all of the activities are complete. All of the activities in the longest path cannot be done until that amount of time has passed.

Go back to page 152 and select another answer.

3.4d. **Correct.** This is the probability of less than 4 accidents. The diagram helps.

$$P(0) + P(1) + P(2) + P(3) + P(4) + \cdots + P(1000) = 1$$

$$P(<4) = P(0) + P(1) + P(2) + P(3)$$

$P(0)$ = solution from Problem 3.2b, page 169 = .135064522

$P(1)$ = solution from Problem 3.1d, page 244 = .270670386

$P(2)$ = solution from Problem 3.1c, page 239 = .270941598

$P(3)$ = solution from Problem 3.4b, page 210 = .180627732

$$P(<4) = P(0) + P(1) + P(2) + P(3)$$

$$= .135064522 + .270670386 + .270941598 + .180627732$$

$$= .857304238$$

Go to the next problem on page 31.

9.2a. **Incorrect.** This is the probability of both 1 and 2 failing at the same time. Think about what must happen in order for a series system to work.

$$P(f)_1 * P(f)_2 = .1 * .2 = .02$$

Go back to page 90 and select another answer.

2.14a. **Incorrect.** This is the probability of no bad resistors. You need to perform a couple more steps.

$$P(0) = \frac{\binom{10}{0}\binom{990}{6}}{\binom{1000}{6}} = \frac{1 * 1.287912 \text{ E15}}{1.368173 \text{ E15}}$$

$$= .941336984$$

Go back to page 22 and select another answer.

14.8a. Correct. LS is simply LF – Dur. As a check, remember that ES and LS should be equal on the critical path. Also, LS – ES should equal S. The table you should generate at this point is

Activity	Dur.	ES	EF	LS	LF	S
a.	12	0	12	0	12	0
b.	3	0	3	4	7	4
c.	5	3	8	7	12	4
d.	2	3	5	13	15	10
e.	2	12	14	20	22	8
f.	10	12	22	12	22	0
g.	7	5	12	15	22	10
h.	5	22	27	22	27	0

Go to the next problem on page 152.

6.1d. Incorrect. This is the correct answer for any failures. If you got this answer, you only need to undo one step to obtain the correct answer.

Go back to page 48 and select another answer.

8.7b. Incorrect. The degrees of freedom are correct, but the error rate is for a one-tailed test. This is $\chi^2_{.1;18}$.

Go back to page 71 and select another answer.

10.9d. Correct. This is determined by adding the probabilities of the individual minimum cut-sets in Problem 10.8b.

$$P(f)_{1,2} = .06 * .08 = .0048$$
$$P(f)_{3,6} = .05 * .03 = .0015$$
$$P(f)_{3,7} = .05 * .02 = .0010$$
$$P(f)_{3,8} = .05 * .01 = .0005$$
$$P(f)_{4,5,7} = .09 * .07 * .02 = .0001260$$
$$P(f)_{4,5,8} = .09 * .07 * .01 = \underline{.0000630}$$
$$P(f)_{sys} = \Sigma \text{ of min cut-sets} = .0079890$$

This completes the cut-set sample problems.

8.4b. Correct. The degrees of freedom for any lower bound are equal to $2 * F$. Since $F = 6$, $v = 12$.

Go to the next problem on page 70.

11.9a. **Incorrect.** This is the answer for $F' \cup B$. Read the problem carefully.

Go back to page 116 and select another answer.

13.6e. **Incorrect.** There is one better solution under minimax.

Go back to page 138 and select another answer.

11.20b. **Incorrect.** This answer is made up. Draw the diagram.

Go back to page 118 and select another answer.

12.4a. **Correct.** The AND gate means multiply B, G, and H.

Go to the next problem on page 129.

2.3b. **Incorrect.** This is the probability of both resistors being good.

$$P(\text{good}) * P(\text{good}) = .6 * .6 = .36$$

Go back to page 17 and select another answer.

14.4d. **Incorrect.** Most of these answers are made up. Quit guessing. To solve for EF of an activity, take the EF of its most important predecessor and add that to the new activity's duration. For example,

$$\text{EF}_c = \text{EF}_b + \text{Dur}_c = 3 + 5 = 8$$

Go back to page 152 and select another answer.

2.10b. **Incorrect.** This would be correct if the problem were not dependent. What happens on one draw affects the next draw. This solution assumes independency, which is incorrect.

Go back to page 18 and select another answer.

11.21d. Correct. Remember that the union of the universe with anything is still the universe. Then the intersection of the universe with anything is just the anything.

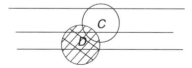

$$P(w \cup C)\, P(D) = 1 * P(D) = 1 * .05$$

This completes the Boolean algebra sample problems.

5.2c. Incorrect. This answer is made up. What do you need to answer the question? How many dust or mist cartridges are needed?

Go back to page 39 and select another answer.

12.1a. Incorrect. There are two kinds of gates—OR and AND. Which kind of gate is under the fault?

Go back to page 128 and select another answer.

9.12c. Incorrect. This is the probability of no accidents for the wrong t. Be careful when you transpose numbers.

$$P(0) = e^{-t/m} = e^{-100/2.3E5} = e^{-4.34782609E-4}$$
$$= .999565312$$

Go back to page 92 and select another answer.

14.6b. Incorrect. At least S_e is incorrect. To compute S subtract EF from LF:

$$S_e = LF_e - EF_e = 22 - 14 = 8$$

Go back to page 152 and select another answer.

10.6d. Incorrect. All of the sets are minimum cut-sets except 3,4,6.

Go back to page 100 and select another answer.

3.6a. Incorrect. This solution is correct for 100 parts in the bin. However, you do not know the number of parts in the bin and cannot assume a number. Therefore, based on one of two assumptions, this problem is binomial—so many parts it is close to independent, or someone is replacing them as fast as you use one.

$$P(1) = \frac{\binom{2}{1}\binom{98}{7}}{\binom{100}{8}} = \frac{\frac{2!}{1!\,1!}\frac{98!}{7!\,91!}}{\frac{100!}{8!\,92!}}$$

$$= \frac{2\,\dfrac{98*97*96*95*94*93*92*91!}{7*6*5*4*3*2*1*91!}}{\dfrac{100*99*98*97*96*95*94*93*92!}{8*7*6*5*4*3*2*1*92!}}$$

$$= \frac{2\,\dfrac{6.97254423\,\text{E}13}{5040}}{\dfrac{7.5030639\,\text{E}15}{40,320}} = \frac{2*13,834,413,152}{186,087,894,300}$$

$$= \frac{27,668,826,304}{186,087,894,300} = .148686869$$

Go back to page 31 and select another answer.

11.11d. Incorrect. This answer is for $P(A \cap B) \cup E$. Be careful in deciding what you want.

Go back to page 116 and select another answer.

10.9c. Incorrect. This is the probability for answer 10.8d.

Go back to page 101 and select another answer.

8.7a. Incorrect. This value is for the incorrect error rate and degrees of freedom. This is a two-tailed upper test. What does the formula require for this bound? This is $\chi^2_{.1;16}$.

Go back to page 71 and select another answer.

12.3d. Incorrect. How can $F2 = F21 + F2$? Examine your gates more closely.

Go back to page 128 and select another answer.

5.2b. **Incorrect.** This solution is for independence. Any probability involving selecting something from a group of finite objects should initially be thought of as dependent.

$$P(0,0,9)\text{ind} = \frac{9!}{0!\,0!\,9!}(.075)^0(.125)^0(.8)^9$$
$$= .8^9 = .134217728$$

Go back to page 39 and select another answer.

7.2c. **Incorrect.** This is the probability of failure between $\bar{x}$ and x.

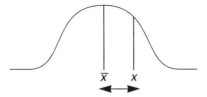

Go back to page 61 and select another answer.

9.2b. **Incorrect.** This is the probability of block 1 working and block 2 failing. Think about what is needed.

$$P(s)_1 * P(f)_2 = .9 * .2 = .18$$

Go back to page 90 and select another answer.

13.7a. **Incorrect.** There is no justification for selecting this alternative under any of the gaming theory techniques discussed. With the possibility of suffering a –3 loss, there does not appear any reason to select this one.

Go back to page 138 and select another answer.

2.11b. Incorrect. This answer is correct for the probability of no bad smoke detectors.

$$P(0) = \frac{\binom{10}{0}\binom{190}{10}}{\binom{200}{10}} = \frac{1 * 1.327869\ \text{E}16}{2.245100\ \text{E}16}$$

$$= .591452134$$

Go back to page 21 and select another answer.

6.3c. Correct. This is the probability of more than 1 failure. Sketch a graph to determine what to do. Then determine λ or m, and t. Since nothing has changed historically, λ and m remain the same as in Problem 6.1.

$$P(0) + P(1) + P(2) + \ldots + P(\infty) = 1$$
$$t = 700$$
$$P(2) + \ldots + P(\infty) = 1 - (P(0) + P(1))$$
$$P(0) \qquad\quad = e^{-t/m} = e^{-700/800} = e^{-.875} = .4168620$$
$$P(1) \qquad\quad = \frac{(t/m)^r e^{-t/m}}{r!} = \frac{(700/800)^1 e^{-.875}}{1!}$$
$$= .875 * .4168620 = .3647543$$
$$P(0) + P(1) \qquad = .7816163$$
$$1 - (P(0) + P(1)) = 1 - .7816163 = .2183837$$

Go to the next problem on page 48.

11.9b. Correct. The diagram shows that the only place the lines intersect is within B.

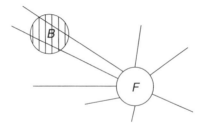

Go to the next problem on page 116.

8.4c. Incorrect. This is correct for upper bounds, but that is not what the question asked. Check the formula again.

Go back to page 70 and select another answer.

14.5a. Incorrect. At least LF_a is incorrect. You must consider the dummy activity as any other activity. Since two activities stem from event 3 (*e* and *f*), when computing latest finish you must use the earliest length of time. This allows all of the activities past that event to remain on schedule. Since event 3 can be complete by 20 (through *e*) or 12 (through *f*), you must use 12. To obtain this correct answer, take LF_f and subtract the duration of *f* from it:

$$LF_a = LF_f - Dur_f = 22 - 10 = 12$$

Go back to page 152 and select another answer.

2.11c. Incorrect. This is the probability of 1 bad resistor.

$$P(1) = \frac{\binom{10}{1}\binom{190}{9}}{\binom{200}{10}} = \frac{10 * 7.3362952\ E14}{2.245100\ E16}$$

$$= .326769135$$

Go back to page 21 and select another answer.

2.6d. Incorrect. This is the probability of exactly 3 good resistors. You were correct to determine this problem is dependent.

$$P(3_{bad}) = \frac{60}{100} * \frac{59}{99} * \frac{58}{98} = \frac{205,320}{970,200} = .211626469$$

Go back to page 18 and select another answer.

14.8c. Incorrect. At least LS_b is incorrect. LS is simply LF − Dur. To obtain the correct answer for LS_b, take LF_b and subtract Dur_b:

$$LS_b = LF_b - Dur_b = 7 - 3 = 4$$

Go back to page 152 and select another answer.

2.9d. **Incorrect.** This is almost correct. You solved it as if it were independent, but it isn't.

Go back to page 18 and select another answer.

12.3c. **Correct.** The mathematical functions for $F21$ and $F22$ are correct, and
the method in which you combined them is correct.

Go to the next problem on page 128.

11.11c. **Incorrect.** This answer is made up. Draw the diagram.

Go back to page 116 and select another answer.

10.9b. **Incorrect.** This is the probability for answer 10.8a.

Go back to page 101 and select another answer.

11.20c. **Correct.** Remember that anything intersecting the universe is simply
itself. Therefore, the union of B and E is simply the addition of the two
since they do not intersect.

$$P(w \cap B) \cup P(E) = P(B) + P(E)$$
$$= .25 + .15 = .4$$

Go to the next problem on page 118.

5.2a. Incorrect. This is the probability of exactly 1 dust, 1 mist, and 7 organic cartridges. The question does not ask for this. Think a little more about how many of each category are required.

$$P(1,1,7) = \frac{\binom{6}{1}\binom{10}{1}\binom{64}{7}}{\binom{80}{9}} = \frac{\frac{6!}{1!5!}\frac{10!}{1!9!}\frac{64!}{7!57!}}{\frac{80!}{9!71!}}$$

$$= \frac{6 * 10 * 621,216,192}{231,900,297,200} = .160728434$$

Go back to page 39 and select another answer.

1.7a. Incorrect. You cannot reduce unlike factorials.

$$\frac{8!}{4!} \neq 2! \qquad \frac{8 * 7 * 6 * 5 * 4 * 3 * 2 * 1}{4 * 3 * 2 * 1}$$

$$= \frac{40,320}{24} \, 2! = 2 * 1 = 2$$

Go back to page 7 and select another answer.

3.1a. Incorrect. This is close, but you found the probability of everything except what you were looking for.

Go back to page 31 and select another answer.

5.4a. Incorrect. This answer is correct for the following setup, which is wrong. Remember your checks to see if the numbers have been correctly inserted into the formula. Also remember to place like categories together.

$$P(1,2) \neq \frac{\binom{6}{2}\binom{14}{1}}{\binom{500}{8}} = \frac{\frac{6!}{2!4!}\frac{14!}{1!13!}}{\frac{500!}{8!492!}}$$

$$= \frac{210}{9.15791275 \text{ E16}} = 2.2930989 \text{ E-}15$$

Go back to page 40 and select another answer.

1.1a. Incorrect. This answer is correct for .004608.

Go back to page 6 and select another answer.

12.9b. Incorrect. If the gate for $F1$ is switched, what is the proper solution for $F1$?

Go back to page 129 and select another answer.

6.1a. Incorrect. This is the probability of exactly 1 failure.

$$1.25e^{-1.25} = 3.581310 \text{ E-1}$$

Go back to page 48 and select another answer.

8.1a. Incorrect. Remember that confidence plus error must equal 1.

Go back to page 70 and select another answer.

2.11d. Correct. This is the probability of any bad resistors. It is derived by:

$$P(0) + P(1) + P(2) + \ldots + P(10) = 1$$
$$P(\text{any}) = 1 - P(0)$$

$$P(0) = \frac{\binom{10}{0}\binom{190}{10}}{\binom{200}{10}} = \frac{1 * 1.327869 \text{ E16}}{2.245100 \text{ E16}}$$

$$= .59145213392$$
$$1 - P(0) = 1 - .59145213392 = .40854786608$$

Go to the next problem on page 21.

9.3d. Incorrect. This is the probability of everything except blocks 3 and 4 working simultaneously.

$$1 - P(s)_3 * P(s)_4 = 1 - .7 * .7 = .51$$

Go back to page 90 and select another answer.

7.1a. Incorrect. This answer is correct for <1380 and >1420.

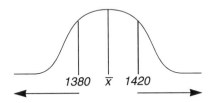

Go back to page 61 and select another answer.

13.1c. Incorrect. If you are looking for the minimum number in each row, you made a mistake. Check again.

Go back to page 137 and select another answer.

12.12b. Incorrect. One mistake is that you have a term that can be reduced. Is it necessary to have *BCR* in the Boolean equivalent? Can it be a minimum cut-set if *B* is a cut-set? See if there is another mistake.

Go back to page 130 and select another answer.

11.1a. Incorrect. This answer is made up. Draw a diagram of the groups needed. Where are the lines?

Go back to page 115 and select another answer.

14.4c. Correct. The diagram shows all of the correct EF's. To compute EF take the EF of its most important predecessor and add that to the new activity's duration. Most important predecessor means that if more than one arrow leads to the event preceding the activity of interest, select the highest EF, because that event cannot actually occur until all activities leading to it are complete, and the next activity cannot begin until that event occurs. The network is

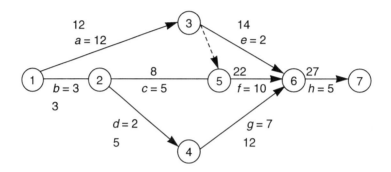

Go to the next problem on page 152.

9.14b. Incorrect. This probability is for no accident for the wrong m. Be careful when you transpose numbers.

$$P(0) = e^{-t/m} = e^{-75/1.4E3} = e^{-5.35714286E\text{-}2}$$
$$= .947838236$$

Go back to page 92 and select another answer.

13.4c. Incorrect. This only yields an expected value of 6. There is a more profitable answer.

Go back to page 138 and select another answer.

11.13b. Correct. Once you realize that A and E do not intersect, all you have left is C.

Go to the next problem on page 117.

12.13d. Correct. This answer can be derived by:

$$F = \cancel{AZ} + \cancel{AE} + \cancel{BC} + \cancel{ADZ} + \cancel{ADE} + \cancel{BCE} + \cancel{BCEZ} + \cancel{ABDEH} + \cancel{HG} + \cancel{BEZ} + \cancel{ABDE} + \cancel{CEZ}$$
$$F = AZ(1+D) + AE + BC + ADE + BCE + BCEZ + ABDEH + HG + BEZ + ABDE + CEZ$$
$$F = AZ + AE(1+D+BDH+BD) + BC + BCE + BCEZ + HG + BEZ + CEZ$$
$$F = AZ + AE + BC(1 + E + EZ) + HG + BEZ + CEZ$$
$$F = AZ + AE + BC + HG + BEZ + CEZ$$

Go to the next problem on page 130.

11.19b. Incorrect. This answer is made up. Draw the diagram. Where are the lines?

Go back to page 117 and select another answer.

14.2c. Correct. By following the arrows in the direction in which they point, trace the path from one event to the next. Remember that dummy activities are treated as other activities. These are the four paths through the diagram.

Go to the next problem on page 152.

11.7d. Incorrect. This is the intersection of B and C, which is needed, but is not the complete answer. Draw the diagram.

Go back to page 116 and select another answer.

1.1b. **Incorrect.** Trick answer! 4608 could be written like this, but for what reason? Scientific notation always has *only* one digit to the left of the decimal place.

Go back to page 6 and select another answer.

8.12a. **Incorrect.** This answer is made up. Take what you know from the problem and determine what unknowns you can solve.

Go back to page 71 and select another answer.

1.6a. **Correct.** Use a calculator, or by hand:

$$5! = 5 * 4 * 3 * 2 * 1 = 120$$

Go to the next problem on page 7.

11.21c. **Incorrect.** What is the intersection of the universe with anything? Draw the diagram.

Go back to page 118 and select another answer.

12.9a. **Correct.** The following drawing represents the dual of the original tree. The bars above the terms represent "not." Once you have the correct dual, derive this answer as follows:

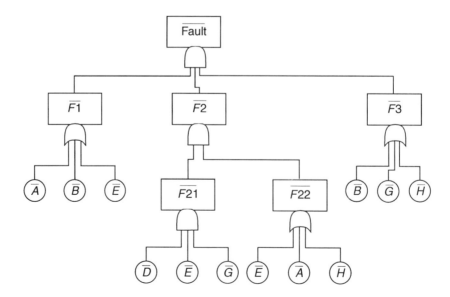

$$
\begin{aligned}
F' &= F1'F2'F3' \\
F1' &= A' + B' + E' \\
F2' &= F21'\,F22' \\
F3' &= B' + G' + H' \\
F21' &= D'E'G' \\
F22' &= E' + A' + H' \\
F' &= (A' + B' + E')(D'E'G'(E' + A' + H'))(B' + G' + H')
\end{aligned}
$$

Go to the next problem on page 129.

7.3a. **Incorrect.** This is the probability of failure between $\bar{x}$ and either one of the x's.

Go back to page 61 and select another answer.

1.12b. **Incorrect.** Quit guessing.

Go back to page 7 and select another answer.

10.5c. **Correct.** Although they are not all minimum cut-sets, each set will cut the system.

Go to the next problem on page 100.

9.10c. **Incorrect.** This probability is for no accident. How can you use it to get the correct answer?

Go back to page 91 and select another answer.

6.5c. **Incorrect.** This answer is correct for exactly 1 failure. Draw the picture to help determine what is needed.

$$
P(1) = \frac{(t/m)^r e^{-t/m}}{r!} = \frac{(600/900)^1 e^{-.6666666}}{1!}
$$
$$
= .6666666 * .5134171 = .3422781
$$

Go back to page 49 and select another answer.

1.1d. **Incorrect.** There are two mistakes here. Scientific notation always has only one digit to the left of the decimal point. The other is that the decimal point was moved in the wrong direction.

Go back to page 6 and select another answer.

8.11d. **Correct.** To obtain this answer you wanted $\chi^2_{.2;34}$. Since the highest number for degrees of freedom for an error of .2 is 30, you had to compute χ^2 from the formula $\chi^2 = .5(z_\alpha + (2v - 1)^{.5})^2$.

$$\begin{aligned}
\chi^2 &= .5(z_\alpha + (2v - 1)^{.5})^2 \\
&= .5(.841 + (2 * 34 - 1)^{.5})^2 \\
&= .5(.841 + (67)^{.5})^2 \\
&= .5(.841 + 8.185353)^2 \\
&= .5(9.026353)^2 \\
&= .5(81.47504) = 40.7375
\end{aligned}$$

$$\lambda < \frac{40.7375}{2(2000)} < .0101844$$

z_α is found at the bottom of the table for each error value. The χ^2 appears reasonable when compared to the highest value in the table. The value 4 less than 36.250 is 31.795. $36.2 - 31.8 = 4.8$, so you would expect your value to be about 4.8 greater. $36.2 + 4.8 = 41$. Your value is close.

Go to the next problem on page 71.

11.19a. **Incorrect.** This answer is made up. Draw the diagram. Where is the union of B and C? Where are the lines?

Go back to page 171 and select another answer.

3.5c. **Incorrect.** This answer is made up.

Go back to page 31 and select another answer.

13.4b. **Incorrect.** This only yields an expected value of .3.

Go back to page 138 and select another answer.

11.1b. **Incorrect.** This answer is correct for the intersection of A and E, but that is not what we want.

Go back to page 115 and select another answer.

12.13c. **Incorrect.** One mistake is that you have a term that can be reduced. Is it necessary to have BCE in the Boolean equivalent? Isn't it not a minimum cut-set if BC is a cut-set? See if there is another mistake.

Go back to page 130 and select another answer.

9.4a. **Correct.** Once blocks 3 and 4 have been combined into block II, they are in series with block 5. In order for a series network to work, all blocks must work.

$$P(s)_{\mathrm{II}} * P(s)_5 = .91 * .6 = .546$$

The redrawn network is

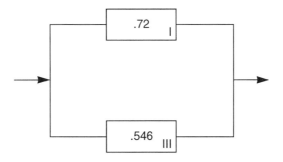

Go to the next problem on page 90.

11.13c. **Incorrect.** This answer is made up.

Go back to page 117 and select another answer.

1.2c. **Incorrect.** You are on the right track. Carefully count the places to the right of the original decimal point.

Go back to page 6 and select another answer.

12.10a. Incorrect. One of these terms is not in the path sets, and there is at least one other mistake. Carefully develop the equations and reduce them to the Boolean equivalent.

Go back to page 129 and select another answer.

9.10b. Correct. For a series network to work, every block in the network must succeed. Since this is an accident system, this is the probability of an accident.

$$P(s)_{\mathrm{I}} * P(s)_{\mathrm{II}} * P(s)_6 = .11536 * .1258 * .10$$
$$= .0014512288$$

Go to the next problem on page 91.

11.7c. Correct. After drawing the diagram there are lines everywhere except the intersection of B and C. Since you want the union, you want everything with lines in it. The easiest way then is to subtract the only area without lines from the universe.

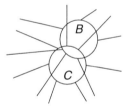

$$P(B' \cup C') = 1 - P(B \cap C) - 1 - (.25 * .1) = .975$$

Go to the next problem on page 116.

9.14c. Incorrect. It appears that you are guessing. This answer could be derived by using the wrong m, but how did you make the exact same mistake?

$$P(f) = 1 - P(0) = 1 - e^{-t/m} = 1 - e^{-75/1.4\mathrm{E}3}$$
$$= 1 - e^{-5.35714286\mathrm{E}\text{-}2} = 1 - .947838236 = .0521617641$$

Go back to page 92 and select another answer.

6.3b. Incorrect. This answer is correct for everything except 1 failure.

$$P(1) \quad = \frac{(t/m)^r e^{-t/m}}{r!} = \frac{(700/800)^1 e^{-.875}}{1!}$$
$$= .875 * .4168620 = .3647543$$
$$1 - P(0) = 1 - .3647543 = .6352457$$

Go back to page 48 and select another answer.

10.5b. Incorrect. All sets are correct except 1,4,7, which will not cut the system.

Go back to page 100 and select another answer.

8.1b. Correct. Confidence plus error must equal 1. Therefore, $1 - CF = \alpha$ or $1 - .9 = .1$.

Go to the next problem on page 70.

13.4a. Incorrect. This is the worst solution under expected value.

Go back to page 138 and select another answer.

3.1c. Correct. This is the probability of exactly 2 accidents. The probability of an accident on any exposure was determined by 4/2000.

$$P(2) = \binom{1000}{2}(.002)^2(.998)^{998}$$

$$= \frac{1000!}{2!\,998!}(4\,\text{E-6})(1.35606406\,\text{E-1})$$

$$= \frac{1000 * 999 * 998!}{2 * 1 * 998!}(5.42425623\,\text{E-7})$$

$$= \frac{999,000}{2}(5.42426\,\text{E-7})$$

$$= 499,500(5.42425623\,\text{E-7}) = .270941598$$

Go to the next problem on page 31.

11.1c. **Incorrect.** This answer is correct for the union of A and F. Be careful when you transpose numbers.

Go back to page 115 and select another answer.

6.7d. **Incorrect.** This answer is made up. Think about what you know. What do you need?

Go back to page 49 and select another answer.

12.13b. **Incorrect.** This is close. At a minimum, you dropped a required cut-set. Review your factoring. It may help to cross out a set after using it.

Go back to page 130 and select another answer.

8.12b. **Correct.** This is computed by working backwards. The problem tells you what you want the upper bound to be (.015). Thus, using the specified $\chi^2_{.05;22}$, the only unknown is T. Solving for T tells you how many tests you would need with no more failures.

$$\frac{\chi^2_{.05;22}}{2T} = .015$$

$$\frac{33.924}{2T} = .015$$

$$33.924 = .015(2T)$$

$$33.924 = .03T$$

$$\frac{33.924}{.03} = T$$

$$T = 1130.8$$

Since you cannot run a partial test, you must round up.

$$1130.8 - 700 = 431$$

This completes the confidence interval sample problems.

12.3a. **Incorrect.** What kind of gate stems from $F2$?

Go back to page 128 and select another answer.

2.4c. **Incorrect.** This answer is almost correct, but you probably made an arithmetic error.

Go back to page 17 and select another answer.

11.18d. **Incorrect.** You might know what you are doing, but you made an arithmetic error.

Go back to page 117 and select another answer.

11.7b. **Incorrect.** This answer is made up.

Go back to page 116 and select another answer.

9.10a. **Incorrect.** This is the probability of blocks I, II, and 6 failing simultaneously. What is needed for a series system to work?

$$P(f)_\mathrm{I} * P(f)_\mathrm{II} * P(f)_6 = .88464 * .8742 * .9$$
$$= .6960170592$$

Go back to page 91 and select another answer.

1.5c. **Incorrect.** The rounding is correct, but check the rules on the power.

Go back to page 6 and select another answer.

12.8c. **Correct.** After you switch all the gates, begin combining them. Since $F2' = F21'\ F22'$, those terms must be multiplied. Since $F21' = D'E'G'$ and $F22' = E' + A' + H'$, $F2' = (D'E'G')(E' + A' + H')$.

Go to the next problem on page 129.

1.3d. **Incorrect.** The rounding is appropriate, but the power is in the wrong direction.

Go back to page 6 and select another answer.

11.13d. **Incorrect.** This answer is made up. Draw the diagram and determine what is left and what is not.

Go back to page 117 and select another answer.

3.5d. **Incorrect.** This answer is made up.

Go back to page 31 and select another answer.

1.12d. **Incorrect.** If you were trying to remember the rule, anything over 0 is not equal to itself.

Go back to page 7 and select another answer.

9.4b. **Incorrect.** This is the probability of block II working and block 5 failing. Think about what is needed.

$$P(s)_{II} * P(f)_5 = .91 * .4 = .364$$

Go back to page 90 and select another answer.

10.5a. **Incorrect.** All sets are correct except for 4,5,6, which will not cut the system.

Go back to page 100 and select another answer.

13.3e. **Incorrect.** At least one of the expected values is incorrect. Pay attention to the negative signs. Adding a negative number is the same as subtracting.

Go back to page 137 and select another answer.

6.1c. **Correct.** The first thing is to determine either λ or m. $\lambda = 15/12000$ (.00125) and $m = 12000/15$ (800). Since $t = 1000$, the probability of no failures is

$$e^{-.00125*1000} = e^{-1000/800} = e^{-1.25} = .2865048$$

Go to the next problem on page 48.

11.10a. **Incorrect.** This answer for $P(C') \cup P(D')$. The question calls for the intersection.

Go back to page 116 and select another answer.

7.3b. Incorrect. This is the probability of failure >1300 or <1500.

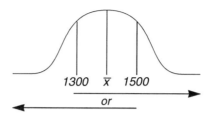

Go back to page 61 and select another answer.

4.1a. Incorrect. This answer was obtained by failing to divide by 2 in the first set of factorials.

$$P(0,1,2,47) = \frac{50!}{0!\,1!\,\textcircled{2!}\,47!}(.001)^0(.014)^1(.025)^2(.96)^{47}$$
$$= .151064817$$

Go back to page 35 and select another answer.

14.5c. Incorrect. LF_h is incorrect. LF_h must equal the latest time that event 7 can occur and be on schedule. Since the critical path determined the program would take 27 days, LF_h must equal 27. If LF_h was correct, then the other LF's would be correct.

Go back to page 152 and select another answer.

9.14d. Incorrect. When the MTBF is presented in network notation, $P(f)$ represents the probability of success for an accident system. Therefore, this is one step too many for this problem.

$$P(f) = 1 - P(0) = 1 - e^{-t/m} = 1 - e^{-75/1.4E2}$$
$$= 1 - e^{-5.35714286E-1} = 1 - .585251104 = .414748896$$

Go back to page 92 and select another answer.

3.1d. **Incorrect.** This is the probability of exactly 1 accident.

$$P(1) = \binom{1000}{1}(.002)^1(.998)^{999}$$

$$= \frac{1000!}{1!\,999!}(.002)(1.35335193\ \text{E}\,\text{-}\,1)$$

$$= \frac{1000 * 999!}{1 * 999!}(2.70670386\ \text{E}\,\text{-}\,4) = 1000(2.70670386\ \text{E}\,\text{-}\,4)$$

$$= .270670386$$

Go back to page 31 and select another answer.

11.1d. **Correct.** This probability is for the union of A and E. Diagram the two groups. Remember that the diagram only has to be accurate to the degree of showing intersections and mutual exclusiveness. Scale and actual position are irrelevant. Draw lines in A and in E. Since this is a union, you want everything with lines in it. They do not intersect, so add the two probabilities.

$$P(A \cup E) = .11 + .15 = .26$$

Go to the next problem on page 115.

12.13a. **Incorrect.** Close. At a minimum, you dropped a required cut-set. Review your factoring. It may help to cross out a set after using it.

Go back to page 130 and select another answer.

1.4d. **Incorrect.** The rounding is incorrect.

Go back to page 6 and select another answer.

11.18c. **Incorrect.** You might know what you are doing, but you made an arithmetic error.

Go back to page 117 and select another answer.

9.12b. Correct. When the MTBF is presented in the network notation, $P(f)$ represents the probability of success for an accident system. Therefore, solve for $P(0)$ and subtract from 1.

$$P(f) = 1 - P(0) = 1 - e^{-t/m} = 1 - e^{-200/2.3E5}$$
$$= 1 - e^{-8.695652E-4} = 1 - .999130813 = 8.69187255 \text{ E-4}$$

Go to the next problem on page 92.

11.7a. Incorrect. This answer is for $B \cup C$, but the problem asks for $B' \cup C'$. Draw the diagram.

Go back to page 116 and select another answer.

12.8b. Incorrect. Close, but every gate must be switched. What would the gate below $F3$ become?

Go back to page 129 and select another answer.

9.9d. Incorrect. You seem to be guessing. This is the probability of block 4 failing and block 5 working. What is needed to solve this problem?

$$P(f)_4 * P(s)_5 = .94 * .07 = .0658$$

Go back to page 91 and select another answer.

8.11b. Incorrect. This answer has a mistake in computing the degrees of freedom (32 instead of 34) at first. The formula $\chi^2 = .5(z_\alpha + (2v - 1)^{.5})^2$ is required, but you used 32 for $2v$ instead of 64. You should have caught this if you did the check discussed in the chapter, since the computed χ^2 of 20.5361 is considerably less than the highest table value of 36.25. Try the correct degrees of freedom in the formula.

Go back to page 71 and select another answer.

7.6d. Incorrect. This answer is made up.

Go back to page 62 and select another answer.

13.3d. Correct. To determine the expected value of each alternative, multiply the probability of each event by the value of the alternative. Then add them.

Go back to page 137 and select another answer.

10.4d. Incorrect. The simultaneous failure of 3 and 7 will cut the system, and no other smaller cut-set contains both blocks.

Go back to page 100 and select another answer.

1.10c. Incorrect. This solution is for a combination. Since order matters, you must use the formula for permutations.

Go back to page 7 and select another answer.

14.3b. Correct. This path takes 27 days. Paths *aeh*, *bcfh*, and *bdgh*, take 19, 23, and 17 days, respectively. The critical path is the longest path and means that each activity on that path must be completed on time (schedule) or the entire program will slip (lengthen or run over schedule).

Go to the next problem on page 152.

8.1c. Incorrect. Remember that confidence plus error must equal 1.

Go back to page 70 and select another answer.

9.4c. Incorrect. This is the probability of failure of the series.

$$1 - P(s)_{\text{II}} * P(s)_5 = 1 - .91 * .6 = .454$$

Go back to page 90 and select another answer.

6.7c. **Correct.** You are trying to determine m, given a desired $P(0)$ in 100 hours. You can use the Poisson distribution where all variables with the exception of m are known. After obtaining the answer, use the fraction or round up. You cannot round down because that does not meet the $P(0) = .25$ requirement.

$$P(0) = e^{-t/m}$$

$$.25 = e^{-100/m}$$

$$\ln(.25) = \frac{-100}{m}\ln e$$

$$-1.38629436112 = \frac{-100}{m}$$

$$-1.38629436112 * m = -100$$

$$m = \frac{-100}{-1.38629436112}$$

$$m = 72.1347520$$

This completes the Poisson sample problems.

11.14a. **Incorrect.** This answer is correct for $(C \cap D) \cup F$.

Go back to page 117 and select another answer.

1.5b. **Incorrect.** Review the placement of the decimal point and the determination of power.

Go back to page 6 and select another answer.

12.14a. **Incorrect.** At least one path set has been omitted from this answer. Carefully reduce the equation through factoring.

Go back to page 130 and select another answer.

4.1b. Correct. This answer is correct for the probability of exactly 0 fatalities, 1 major accident, and 2 minor accidents. There must be 47 safe trips to account for all trips.

$$P(0,1,2,47) = \frac{50!}{0!\,1!\,2!\,47!}(.001)^0(.014)^1(.025)^2(.96)^{47}$$

$$= \frac{50 * 49 * 48 * 47!}{1 * 1 * 2 * 1 * 47!}(1)(.014)(.000625)(1.46807402\,E\text{-}1)$$

$$= \frac{117,600}{2}1.28456477\,E\text{-}6 = 58,800 * 1.28456477\,E\text{-}6$$

$$= .0755324083$$

Go to the next problem on page 35.

13.1a. Incorrect. Under a maximin strategy, which numbers do you circle? Remember why it is called maximin.

Go back to page 137 and select another answer.

7.3c. Incorrect. This is the probability of failure <1300 or >1500.

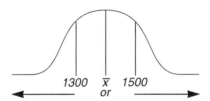

Go back to page 61 and select another answer.

12.12d. Incorrect. Close. Somehow, you dropped a cut-set that is necessary. Review your factoring. It may help to cross out a set after using it.

Go back to page 130 and select another answer.

1.11d. Incorrect. How did you derive this number?

Go back to page 7 and select another answer.

2.9a. **Correct.** This is the probability of 2 or more resistors in 3 draws, assuming dependency. The picture helps to see the solution.

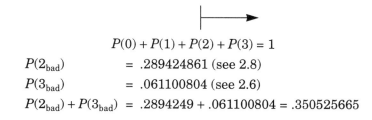

$$P(0) + P(1) + P(2) + P(3) = 1$$

$P(2_{bad})$ $\quad = .289424861$ (see 2.8)

$P(3_{bad})$ $\quad = .061100804$ (see 2.6)

$P(2_{bad}) + P(3_{bad}) \ = .2894249 + .061100804 = .350525665$

Go to the next problem on page 18.

9.9c. **Incorrect.** It appears that you are guessing. This is the probability of block 4 working and block 5 failing. What is needed to solve this problem?

$$P(s)_4 * P(f)_5 = .06 * .93 = .0558$$

Go to the next problem on page 91.

11.18b. **Correct.** After drawing the diagram and filling in the lines, the only place where B intersects C and they intersect D is the small intersection of all three circles. Compute it by multiplying all three probabilities, since they are independent.

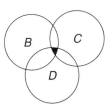

$$P(B \cap C) \cap P(D) \ = P(B) * P(C) * P(D)$$
$$= .25 * .1 * .05 = .00125$$

Go to the next problem on page 117.

6.5b. **Incorrect.** This answer is made up. Use what you can determine—m and t.

Go back to page 49 and select another answer.

8.12c. **Incorrect.** This answer is made up. Think a little.

Go back to page 71 and select another answer.

2.4b. **Correct.** This is the answer for 2 bad resistors.

$$P(\text{bad}) * P(\text{bad}) = .4 * .4 = .1600$$

Go to the next problem on page 17.

11.2a. **Incorrect.** Close, but you forgot to subtract the intersection. Draw the picture and you should see that you included the intersection twice.

Go back to page 115 and select another answer.

13.3c. **Incorrect.** At least one expected value is incorrect. Pay attention to the negative signs. Adding a negative number is the same as subtracting.

Go back to page 137 and select another answer.

1.5d. **Correct.** By now I think you have the idea. A 5 is rounded to the nearest even number, which causes the 9 to become 0 and the 8 to become a 9. Power is determined by the distance and direction from the original decimal.

Go to the next problem on page 67.

12.8a. **Incorrect.** The dual takes every gate and switches it. Therefore, the OR gate for F in the original must become an AND gate.

Go back to page 129 and select another answer.

11.6d. **Incorrect.** This answer is made up. Draw the diagram. Where is A'? F'? Where are the lines?

Go back to page 116 and select another answer.

10.4c. **Correct.** The simultaneous failure of 3 and 6 alone will cut the system, so it is not necessary to include 4 as part of the cut-set.

Go to the next problem on page 100.

7.1b. **Incorrect.** This answer is correct for >1380.

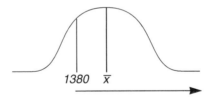

Go back to page 61 and select another answer.

9.15a. **Incorrect.** This is the probability of success for a block in a reliability system where MTBF data is provided. Use the Poisson distribution to determine $P(0)$. The question asks for failure of the block.

$$P(0) = e^{-t/m} = e^{-80/4.5E3} = e^{-1.7777778E-2}$$
$$= .982379315$$

Go back to page 93 and select another answer.

2.2c. **Incorrect.** This answer is incorrect for two reasons. First it appears that you used the same logic as 2.2a, page 173. The second reason is that there is an arithmetic mistake ($.6 * .4 \neq .02400$).

Go back to page 17 and select another answer.

12.9d. **Incorrect.** What kind of gate is below the top-level fault in the dual?

Go back to page 129 and select another answer.

8.11a. **Incorrect.** This answer was derived with one major error. The degrees of freedom were computed not with $2F + 2$, but F. Thus you found $\chi^2_{.2;16}$, which is not correct. Determine the correct degrees of freedom.

Go back to page 71 and select another answer.

1.6b. **Incorrect.** This number is made up.

Go back to page 7 and select another answer.

12.14b. Correct. The cut-sets can be written mathematically as the formula for the Boolean equivalent:

$$F = BE + DE + DJ + EGJ + GHIJ$$

Therefore, the path sets' equation is

$$F' = (B' + E')(D' + E')(D' + J')(E' + G' + J')(G' + H' + I' + J')$$

The above was obtained by reversing all of the gates in the Boolean equivalent. Proceeding,

$$F' = (B'D' + B'E' + D'E' + E')(D' + J')(E' + G' + J')(G' + H' + I' + J')$$
$$F' = (B'D' + E'(B' + D' + 1))(D' + J')(E' + G' + J')(G' + H' + I' + J')$$
$$F' = (B'D' + E')(D' + J')(E' + G' + J')(G' + H' + I' + J')$$
$$F' = (B'D' + B'D'J' + D'E' + E'J')(E' + G' + J')(G' + H' + I' + J')$$
$$F' = (B'D'(1 + J') + D'E' + E'J')(E' + G' + J')(G' + H' + I' + J')$$
$$F' = (B'D' + D'E' + E'J')(E' + G' + J')(G' + H' + I' + J')$$
$$F' = (B'D'E' + B'D'G' + B'D'J' + D'E' + D'E'G' + D'E'J' + E'J' + E'G'J' + E'J')(G' + H' + I' + J')$$
$$F' = (B'D'G' + B'D'J' + D'E'(B' + 1 + G' + J') + E'J' + E'G'J' + E'J')(G' + H' + I' + J')$$
$$F' = (B'D'G' + B'D'J' + D'E' + E'J'(1 + G' + 1))(G' + H' + I' + J')$$
$$F' = (B'D'G' + B'D'J' + D'E' + E'J')(G' + H' + I' + J')$$
$$F' = B'D'G' + B'D'G'H' + B'D'G'I' + B'D'G'J' + B'D'G'J' + B'D'H'J' + B'D'I'J' + B'D'J' + D'E'G' + D'E'H' + D'E'I' + D'E'J' + E'G'J' + E'H'J' + E'I'J' + E'J'$$
$$F' = B'D'G'(1 + H' + I' + J' + J') + B'D'H'J' + B'D'I'J' + B'D'J' + D'E'G' + D'E'H' + D'E'I' + D'E'J' + E'G'J' + E'H'J' + E'I'J' + E'J'$$
$$F' = B'D'G' + B'D'J'(H' + I' + 1) + D'E'G' + D'E'H' + D'E'I' + D'E'J' + E'G'J' + E'H'J' + E'I'J' + E'J'$$
$$F' = B'D'G' + B'D'J' + D'E'G' + D'E'H' + D'E'I' + E'J'(D' + G' + H' + I' + 1)$$
$$F' = B'D'G' + B'D'J' + D'E'G' + D'E'H' + D'E'I' + E'J'$$

This completes the fault-tree analysis sample problems.

13.1b. Incorrect. If you are looking for the minimum number in each row, you made a mistake. Check again.

Go back to page 137 and select another answer.

4.1c. **Incorrect.** This answer would be correct except for a mistake in scientific notation.

Go back to page 35 and select another answer.

12.12c. **Incorrect.** One mistake is that a term can be reduced. Is it necessary to have BC in the Boolean equivalent? Can it be a minimum cut-set if B is a cut-set? See if there is another mistake.

Go back to page 130 and select another answer.

6.7b. **Incorrect.** This answer is close, and you may know what you are doing. However, the MTBF must equal or exceed the number in order for $P(0) = .25$.

Go back to page 49 and select another answer.

13.3b. **Incorrect.** At least one expected value is incorrect. Pay attention to the negative signs. Adding a negative number is the same as subtracting.

Go back to page 137 and select another answer.

11.14b. **Incorrect.** This answer is made up. Draw the diagram and try again.

Go back to page 117 and select another answer.

8.1d. **Incorrect.** Remember that confidence plus error must equal 1.

Go back to page 70 and select another answer.

9.4d. **Incorrect.** This is the probability of blocks II and 5 failing. Think about what must happen in series.

$$P(f)_{\mathrm{II}} * P(f)_5 = .09 * .4 = .036$$

Go back to page 90 and select another answer.

11.18a. **Incorrect.** This answer is made up. Draw the diagram. Where does the intersection of B and C intersect D?

Go back to page 117 and select another answer.

1.2d. **Correct.** You appropriately determined the placement of the decimal point (four places to the right of the original decimal point).

Go to the next problem on page 6.

10.4b. **Incorrect.** The simultaneous failure of 4, 5, and 8 will cut the system, and no smaller cut-set contains all three blocks.

Go back to page 100 and select another answer.

3.2a. **Incorrect.** This is not possible since it is greater than 1. You may have copied something wrong.

Go back to page 31 and select another answer.

9.9b. **Incorrect.** This is the probability of failure for Part II. You are close. Think about the next step.

$$P(f)_4 * P(f)_5 = .94 * .93 = .8742$$

Go back to page 91 and select another answer.

12.7d. **Incorrect.** This number is made up. Quit guessing. What do you need to do before substituting numbers into the equation?

Go back to page 129 and select another answer.

1.7b. **Correct.** It could be computed by calculator or by hand.

$$\frac{8!}{4!} = \frac{8 * 7 * 6 * 5 * 4!}{4!} = 1680$$

Go to the next problem on page 7.

11.6c. **Incorrect.** This is the correct answer for the union of A and F. What is the question?

Go back to page 116 and select another answer.

1.4c. **Incorrect.** The power must be negative.

Go back to page 6 and select another answer.

8.10d. Incorrect. This answer was obtained with the correct χ^2 but without multiplying T by 2 in the denominator. Be more careful, especially if you get the hard part—the χ^2.

Go back to page 71 and select another answer.

1.8a. Incorrect. This answer is for 9^2, but that has nothing to do with this problem.

Go back to page 7 and select another answer.

12.14c. Incorrect. At least one extra path set should be obvious. If $B'D'$ is a path set, other path sets could be factored.

Go back to page 130 and select another answer.

11.2b. Correct. This is the probability of the union of B and D. Diagram them. Remember that the picture only has to be accurate to the degree of showing intersections and mutual exclusiveness. Scale and actual position are irrelevant. Draw lines in B and D. Since this is a union, you want everything with lines in it. Since they intersect, add the two probabilities and subtract the intersection. There is no indication that they are dependent, so multiply the probabilities to obtain the value for the intersection.

$$P(B \cup D) = P(B) + P(D) - P(B \cap D)$$
$$= .25 + .05 - (.25 * .05) = .3 - .0125 = .2875$$

Go to the next problem on page 115.

13.3a. Incorrect. At least one expected value is incorrect. Pay attention to the negative signs. Adding a negative number is the same as subtracting.

Go back to page 137 and select another answer.

7.3d. Correct. First sketch the graph to determine what is desired. Next, determine the z value. Since one x is to the right of $\bar{x}$ and the other is to the left, you add the z_{rep}'s to determine the area between the two. Since the x's are the same distance from $\bar{x}$, the z_{rep}'s are the same.

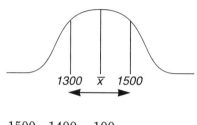

$$z = \frac{1500 - 1400}{50} = \frac{100}{50} = 2.0$$

$$z_{rep} = .47724$$

$$z = \frac{1300 - 1400}{50} = \frac{100}{50} = 2.0$$

$$z_{rep} = .47724$$

$$P(1300 < f < 1500) = .47724 + .47724 = .95448$$

Go to the next problem on page 61.

2.10d. Correct. This is the probability of any bad resistors. It is derived by

$$P(0_{bad}) = \frac{60}{100} * \frac{59}{99} * \frac{58}{98} = \frac{205,320}{970,200} = .2116265$$

$$1 - P(0) = 1 - .211626469 = .7883735$$

Continue the chapter on page 18.

9.9a. Correct. This is the easiest way to solve for success of a parallel system.

$$P(s)_{II} = 1 - P(f)_4 * P(f)_5 = 1 - .94 * .93$$
$$= 1 - .8742 = .1258$$

The network can now be redrawn as

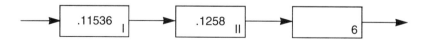

Go to the next problem on page 91.

4.1d. **Incorrect.** The most common mistake here is to forget to account for the 47 safe trips. The formula you are attempting to solve might look like this:

$$P(0,1,2,) = \frac{50!}{0!\,1!\,2!}(.001)^0(.014)^1(.025)^2 = \ldots$$

Go back to page 35 and select another answer.

9.15b. **Incorrect.** This answer is made up. By now you should be able to do this one.

Go back to page 93 and select another answer.

1.8c. **Incorrect.** This is multiplication not division.

Go back to page 7 and select another answer.

11.17d. **Incorrect.** This answer is made up. Draw the diagram.

Go back to page 117 and select another answer.

11.6b. **Incorrect.** This answer is correct for the intersection of A' and F', but that is not what is sought.

Go back to page 116 and select another answer.

10.4a. **Incorrect.** The simultaneous failure of 1 and 2 will cut the system, and no smaller cut-set contains both blocks.

Go back to page 100 and select another answer.

8.12d. **Incorrect.** This answer is made up. Think about what you know and what you don't know.

Go back to page 71 and select another answer.

11.14c. Correct. Although C and D intersect, they do not intersect with F. Therefore, the answer is 0.

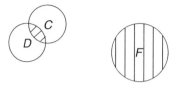

Go to the next problem on page 117.

6.7a. Incorrect. This answer is made up. What do you know? What is $P(0)$?

Go back to page 49 and select another answer.

13.2e. Incorrect. You can do better than guaranteeing that you will lose 2 if the worst alternative for each solution occurs.

Go back to page 137 and select another answer.

1.4a. Incorrect. The power is correct, but the rounding is off.

Go back to page 6 and select another answer.

2.5a. Incorrect. This is the probability of everything except exactly 1 bad resistor.

$$P(1_{bad}) = P(b,g) + P(g,b) = (.4 * .6) + (.6 * .4)$$
$$= .24 + .24 = .4800$$
$$1 - P(1_{bad}) = 1 - .4800 = .5200$$

Go back to page 17 and select another answer.

5.3d. Incorrect. This number is made up. Reread the question and look for things you know. Then try to fit them into the hypergeometric formula because they meet the requirements for that formula.

Go back to page 39 and select another answer.

14.8b. **Incorrect.** At least LS_c is incorrect. LS is simply LF – Dur. To obtain the correct answer for LS_c, take LF_c and subtract Dur_c from it:

$$LS_c = LF_c - Dur_c = 12 - 5 = 7$$

Go back to page 153 and select another answer.

1.1c. **Correct.** To arrive at this answer, count the number of decimal places between the original decimal and the digit farthest to the left. Since you moved the decimal point to the left, E is positive.

Go to the next problem on page 6.

10.7a. **Incorrect.** A minimum cut-set is missing.

Go back to page 100 and select another answer.

2.12a. **Incorrect.** This is close to the correct answer for the probability of any bad capacitors.

Go back to page 21 and select another answer.

9.12a. **Incorrect.** This is the probability of failure of the block. Remember that when the MTBF is used, $P(f)$ represents the probability of success for an accident system. It is much easier in these cases to use a table and label it.

$$P(0) = e^{-t/m} = e^{-200/2.3E5} = e^{-8.695652E-4}$$
$$= .999130813$$

Go back to page 192 and select another answer.

14.6a. Correct. Slack is simply the difference between LF and EF. The critical path can have no slack. That is one check to ensure accuracy of computation. Another is that, along a path, once there is slack it will not change unless there are two activities into the same event. The completed diagram is

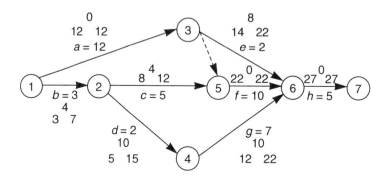

Go to the next problem on page 153.

9.2c. Correct. This is the probability of both blocks working. In series, this is necessary for the system to work.

$$P(s)_1 * P(s)_2 = .9 * .8 = .72$$

The new network is

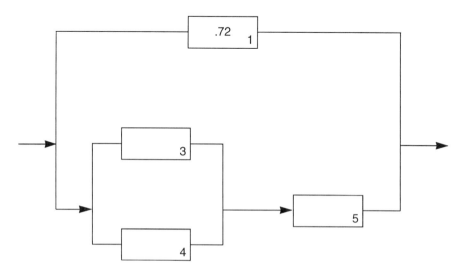

Go to the next problem on page 90.

3.6b. **Incorrect.** This solution is correct for 1000 parts in the bin. However, you do not know the number of parts in the bin and cannot assume a number. Therefore, this is a binomial problem based on one of two assumptions—so many parts that it is close to independent, or someone is replacing them as fast as you use one.

$$P(1) = \frac{\binom{20}{1}\binom{980}{7}}{\binom{1000}{8}} = \frac{\dfrac{20!}{1!\,19!}\dfrac{980!}{7!\,973!}}{\dfrac{1000!}{8!\,992!}}$$

$$= \frac{20\,\dfrac{980 * 979 * 978 * 977 * 976 * 975 * 974 * 973!}{7 * 6 * 5 * 4 * 3 * 2 * 1 * 973!}}{\dfrac{1000 * 999 * 998 * 997 * 996 * 995 * 994 * 993 * 992!}{8 * 7 * 6 * 5 * 4 * 3 * 2 * 1 * 992!}}$$

$$= \frac{20\,\dfrac{8.49680353\text{ E}20}{5040}}{\dfrac{9.72320047\text{ E}23}{40,320}} = \frac{20 * 1.68587372\text{ E}17}{2.41150805\text{ E}19}$$

$$= \frac{3.37174743\text{ E}18}{2.41150805\text{ E}19}$$

$$= .139819041$$

Go back to page 31 and select another answer.

8.6d. **Incorrect.** This answer uses 2 times the degrees of freedom. This is $\chi^2_{.05;24}$.

Go back to page 70 and select another answer.

7.2d. **Incorrect.** This number is made up.

Go back to page 61 and select another answer.

2.12b. Correct. This is the probability of all good capacitors or no bad ones. The total in the numerator was determined by adding the two categories.

$$P(0) = \frac{\binom{4}{0}\binom{146}{5}}{\binom{150}{5}} = \frac{1 * 515{,}853{,}624}{591{,}600{,}030}$$

$$= .871963485$$

Go to the next problem on page 22.

6.3d. Incorrect. This answer is correct for the probability of 0 or 1 failures (< 2). It is very close to the correct answer. Draw a picture.

$$P(0) = e^{-t/m} = e^{-700/800} = e^{-.875} = .4168620$$

$$P(1) = \frac{(t/m)^r e^{-t/m}}{r!} = \frac{(700/800)^1 e^{-.875}}{1!}$$

$$= .875 * .4168620 = .3647543$$

$$P(0) + P(1) = .7816163$$

Go back to page 48 and select another answer.

8.4d. Incorrect. The formula says to multiply, not divide, by 2.

Go back to page 70 and select another answer.

14.5b. Correct. To compute the LF's start with the last event. Every activity that comes into that event has an LF equal to when that event can occur. In this case, since the critical path resulted in 27 days, event 7 occurs on day 27 and every activity leading to it (just h here) has an LF of 27. To compute when event 6 must occur, subtract Dur_h from LF_h. Now every activity leading to event 6 must have an LF equal to that number. To treat multiple paths stemming from an event, you must use the lowest number as the LF for all activities coming into the event. Also, LF and EF are equal on the critical path. The diagram including both EF and LF (boxes not shown) is

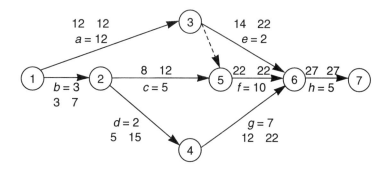

Go to the next problem on page 152.

13.7b. Maybe. Since this alternative was the alternative of choice for both the maximin and minimax strategies, there is ample justification for this solution, especially since it covers both the pessimist and the optimist.

Go to the next answer on page 138.

11.9c. Incorrect. This answer is correct for $F \cap B'$. Read the problem carefully.

Go back to page 116 and select another answer.

10.9a. Incorrect. This is the probability of the answer for 10.8c.

Go back to page 101 and select another answer.

14.2d. Incorrect. You select paths by following arrows in the direction in which they point. How did you get from b to c to e? Carefully select each path.

Go back to page 152 and select another answer.

7.4d. Incorrect. This answer is made up.

Go back to page 61 and select another answer.

12.3b. Incorrect. The mathematical functions for $F21$ and $F22$ are incorrect, but the resolution of $F21$ and $F22$ is correct. What kind of gates are below $F21$ and $F22$?

Go back to page 128 and select another answer.

8.6c. Incorrect. There are two mistakes here. It is not necessary to do anything to the degrees of freedom when they are provided in the formula. This is $\chi^2_{.5;24}$.

Go back to page 70 and select another answer.

11.20d. Incorrect. This answer is made up. Draw the diagram. What is the intersection of anything with w?

Go back to page 118 and select another answer.

10.7b. Correct. There are additional cut-sets, but all of the minimum cut-sets are included.

Go to the next problem on page 100.

1.4b. Correct. The power placement and rounding are both correct.

Go to the next problem on page 6.

14.8d. Incorrect. At least LS_e is incorrect. LS is simply LF – Dur. To obtain the correct answer for LS_e, take LF_e and subtract Dur_e from it:

$$LS_e = LF_e - Dur_e = 22 - 2 = 20$$

Go back to page 152 and select another answer.

2.7b. Incorrect. This probability assumes independency, which is incorrect (see 2.7a, page 209). It also contains an arithmetic error.

$$.09600 \neq 9.6\ E\text{-}1$$

Go back to page 18 and select another answer.

6.5d. Incorrect. This answer is made up.

Go back to page 49 and select another answer.

2.7d. Incorrect. You probably know what you are doing, but you made an arithmetic mistake.

Go back to page 18 and select another answer.

2.13a. Correct. This is the probability of having 2 defective hard hats out of 12 drawn from the bin.

$$P(2) = \frac{\binom{5}{2}\binom{495}{10}}{\binom{500}{12}} = \frac{\dfrac{5!}{2!\,3!}\dfrac{495!}{10!\,485!}}{\dfrac{500!}{12!\,488!}}$$

$$= \frac{\dfrac{5*4*3!}{2*3!}\dfrac{495*494*493*492*491*490*489*488*487*486*485!}{10*9*8*7*6*5*4*3*2*485!}}{\dfrac{500*499*498*497*496*495*494*493*492*491*490*489*488!}{12*11*10*9*8*7*6*5*4*3*2*488!}}$$

$$= \frac{\dfrac{20}{2}\dfrac{8.059640\,E26}{3,628,800}}{\dfrac{2.137315\,E32}{479,001,600}} = \frac{10*2.221021\,E20}{4.462021\,E23}$$

$$= 4.97761183\,E\text{-}3$$

Go to the next problem on page 22.

2.4d. Incorrect. This answer is almost correct, but you made a major arithmetic error.

Go back to page 17 and select another answer.

9.11d. Incorrect. This is made up. Think about what you did for Problem 9.10, page 91.

Go back to page 92 and select another answer.

2.1d. Correct. This is the probability of a bad resistor in 1 draw.

$$P(\text{bad}) = 40 \ (\text{bad res})/100 \ (\text{total res}) = .4000$$

Go to the next problem on page 17.

5.1d. Incorrect. This probability is correct for one too many organic cartridges. The totals of the numerator must equal the numbers in the denominator.

$$P(2,4,4) = \frac{\binom{6}{2}\binom{10}{4}\binom{64}{4}}{\binom{80}{9}} = \frac{\dfrac{6!}{2!\,4!}\dfrac{10!}{4!\,6!}\dfrac{64!}{4!\,60!}}{\dfrac{80!}{9!\,71!}}$$

$$= \frac{\dfrac{30}{2}\dfrac{5040}{24}\dfrac{15,249,024}{24}}{\dfrac{8.41519798\ E16}{362,880}}$$

$$= \frac{15 * 210 * 635,376}{231,900,297,200}$$

$$= \frac{2,001,434,400}{231,900,297,200} = 8.63058144\ E\text{-}3$$

Go back to page 39 and select another answer.

11.11b. Incorrect. This answer is made up. Draw the diagram. What intersects?

Go back to page 116 and select another answer.

10.8d. Incorrect. At a minimum, this list contains a set that is not a cut-set (3,4,5) and is missing at least one minimum cut-set.

Go back to page 100 and select another answer.

8.6b. Correct. Find .05 across the top and 12 down the left column.

Go to the next problem on page 71.

14.3a. Incorrect. This path only requires 23 days. There is at least one longer path. Remember, the critical path is the longest path because the job cannot be done until all of the activities are complete. All of the activities in the longest path cannot be done until that amount of time has passed.

Go back to page 152 and select another answer.

9.2d. Incorrect. This is the probability of block 1 failing and block 2 working.

$$P(f)_1 * P(s)_2 = .1 * .8 = .08$$

Go back to page 90 and select another answer.

7.4c. Incorrect. This answer is for the area between $\bar{x}$ and x.

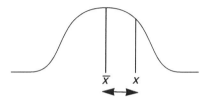

Go back to page 61 and select another answer.

2.13c. Incorrect. This is the probability of any defective hard hats.

$$1 - P(0) = P(\text{any})$$

$$P(0) = \frac{\binom{5}{0}\binom{495}{12}}{\binom{500}{12}} = \frac{1 * 3.94972 \text{ E23}}{4.462021 \text{ E23}}$$

$$= .885185303$$
$$1 - P(0) = 1 - .885185303 = .114814697$$

Go back to page 22 and select another answer.

2.9b. Incorrect. This solution is only for part of the problem. Try drawing a picture to help figure out what you need.

Go back to page 18 and select another answer.

2.6b. Incorrect. This answer would be correct if the problem were not dependent. What happens on one draw affects the next draw.

$$P(3_{bad(ind)}) = .4 * .4 * .4 = .06400$$

Go back to page 18 and select another answer.

3.6c. Correct. This is the same as asking for the probability of 1 bad part. It is independent because there is not enough information to solve it as a dependent problem. Therefore, you assume that there is either replacement, or so many parts that it acts as if it were independent.

$$P(1) = \binom{8}{1}(.02)^1(.98)^7 = \frac{8!}{1!7!}(.02)(.868125533)$$

$$= \frac{8 * 7!}{1 * 7!}(.0173625107)$$

$$= .138900085$$

This completes the binomial sample problems.

13.7c. Incorrect. There is no real justification for selecting this alternative. Are you serious or just guessing?

Go back to page 138 and select another answer.

2.13d. Incorrect. This is the probability of no defective hats. See Problem 2.13c, page 267.

Go back to page 22 and select another answer.

14.9a. Incorrect. H is on the critical path; therefore, it has no extra resources. Think about which activities have extra resources (slack).

Go back to page 153 and select another answer.

9.11c. Incorrect. This is the probability that all of the major blocks fail. See 9.10a, page 241.

Go back to page 92 and select another answer.

5.1c. **Incorrect.** This answer is made up. Take a moment and think. Is the problem independent or dependent? Is it two-cased or multiple-cased? Which formula is needed for this type problem?

Go back to page 39 and select another answer.

11.9d. **Incorrect.** This answer is made up. Draw the diagram. Where do lines intersect?

Go back to page 116 and select another answer.

10.7c. **Incorrect.** One minimum cut-set is missing.

Go back to page 100 and select another answer.

11.21a. **Incorrect.** This answer is made up. Draw the diagram and determine what is needed.

Go back to page 118 and select another answer.

2.5c. **Incorrect.** This answer is for exactly 2 good resistors. It is needed for the correct answer, but is not correct by itself.

$$P(2good) = .6 * .6 = .3600$$

Go back to page 17 and select another answer.

11.11a. **Correct.** Draw the union of A and B. Then draw E and note that since you want the union of it with A and B, you want everything with a line in it. There are not intersections, so you need not subtract anything.

$$P(A \cup B) \cup C = P(A) + P(B) + P(E)$$
$$= .11 + .25 + .15 = .51$$

Go to the next problem on page 116.

8.6a. Incorrect. Be careful when you use the table. Converting from percentages into decimals is sometimes tricky if you rush. This is the value for an error rate of .5.

Go back to page 70 and select another answer.

14.2b. Incorrect. Select paths by following the arrows in the direction in which each points. How did you get from b to c to f to g? Carefully select each path.

Go back to page 152 and select another answer.

9.11b. Correct. If you are working these in order, hopefully you noticed that you only needed to subtract the answer from 9.10b from 1.

$$P(f) = 1 - P(s)_{\text{I}} * P(s)_{\text{II}} * P(s)_6 = 1 - .11536 * .1258 * .10$$
$$= 1 - .0014512288 = .9985487712$$

Go to the next problem on page 92.

12.2d. Incorrect. This notation is basically meaningless. The closest thing it could mean is to divide A by B and then divide by E, but there is no notation for division in Boolean logic.

Go back to page 128 and select another answer.

1.3b. Incorrect. There are two mistakes. The power is incorrect, and you forgot the rounding rules.

Go back to page 6 and select another answer.

9.3a. Incorrect. This is the probability of 3 and 4 failing at the same time, which is needed to find the correct answer. Think about what must happen for a parallel system to work.

$$P(f)_3 * P(f)_4 = .3 * .3 = .09$$

Go back to page 90 and select another answer.

7.4b. Incorrect. This is the correct answer for $f > 1500$.

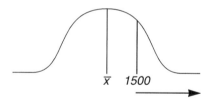

Go back to page 61 and select another answer.

14.6c. Incorrect. At least S_f is incorrect. This should be obvious because f is on the critical path and, by definition, can contain no slack. To compute S subtract EF from LF:

$$S_f = LF_f - EF_f = 22 - 22 = 0$$

Go back to page 152 and select another answer.

1.11b. Correct. The problem calls for the formula for combinations.

$$\binom{27}{3} = \frac{27!}{3!\,(27-3)!}$$

$$= \frac{27 * 26 * 25 * 24!}{3 * 2 * 1 * 24!} = \frac{17,550}{6}$$

$$= 2925$$

Go to the next problem on page 7.

2.3a. Incorrect. This is the probability of getting no good resistors in 2 draws.

$$P(\text{bad}) * P(\text{bad}) = .4 * .4 = .1600 = 1.600 \text{ E-1}$$

Go back to page 17 and select another answer.

2.14d. Correct. This is the probability of less than 3 bad resistors.

$$P(0) + P(1) + P(2) + P(3) + P(4) + \ldots + P(10) = 1$$
$$P(< 3) = P(0) + P(1) + P(2)$$

$$P(0) \quad = \frac{\binom{10}{0}\binom{990}{6}}{\binom{1000}{6}} = \frac{1 * 1.287912\,\text{E}15}{1.368173\,\text{E}15}$$

$$= .941336984$$

$$P(1) \quad = \frac{\binom{10}{1}\binom{990}{5}}{\binom{1000}{6}} = \frac{10 * 7.845150\,\text{E}12}{1.368173\,\text{E}15}$$

$$= .0573403239$$

$$P(2) \quad = \frac{\binom{10}{2}\binom{990}{4}}{\binom{1000}{6}} = \frac{45 * 39,782,707,965}{1.368173\,\text{E}15}$$

$$= 1.30847595\,\text{E-3}$$

$$P(0) + P(1) + P(2) = .941336984 + .0573403239 + 1.30847595\,\text{E-3}$$
$$= .999985784$$

Go to the next problem on page 22.

10.8b. Correct. This is the list of minimum cut-sets.

Go to the next problem on page 101.

11.10d. Incorrect. This answer is made up. Draw the diagram. Where do the lines cross?

Go back to page 116 and select another answer.

5.1b. **Correct.** This solution is for exactly 2 dust, 4 mist, and 3 organic cartridges. The important step is to realize that not all of the cartridges needed are accounted for in the original question. Therefore, you must need the remainder in organic cartridges since there are only three categories.

$$P(2,4,3) = \frac{\binom{6}{2}\binom{10}{4}\binom{64}{3}}{\binom{80}{9}} = \frac{\frac{6!}{2!\,4!}\,\frac{10!}{4!6!}\,\frac{64!}{3!61!}}{\frac{80!}{9!71!}}$$

$$= \frac{\frac{6*5*4!}{2*1*4!}\,\frac{10*9*8*7*6!}{4*3*2*1*6!}\,\frac{64*63*62*61!}{3*2*1*61!}}{\frac{80*79*78*77*76*75*74*73*72*71!}{9*8*7*6*5*4*3*2*1*71!}}$$

$$= \frac{\frac{30}{2}\,\frac{5040}{24}\,\frac{249,984}{6}}{\frac{8.41519798\,E16}{362,880}} = \frac{15*210*41,664}{231,900,297,200}$$

$$= \frac{131,241,600}{231,900,297,200} = 5.65939766\,E\text{-}4$$

Go to the next problem on page 39.

8.5d. **Incorrect.** The formula does not call for division by 2.

Go back to page 70 and select another answer.

4.4d. **Incorrect.** This problem can be solved as a multicategory, independent probability.

This completes the multinomial sample problems.

2.7c. **Correct.** This answer is for the specific order given if the problem is dependent. Since the automatic assumption when taking parts from a parts bin should be dependency, and the information necessary to determine the dependency is available, work it as a dependent problem.

$$P(\text{bad}) * P(\text{good}) * P(\text{bad}) = \frac{40}{100} * \frac{60}{99} * \frac{39}{98}$$

$$= \frac{93,600}{970,200} = .096474954$$

Go to the next problem on page 18.

10.7d. **Incorrect.** This list contains at least one set that is not even a cut-set.

Go back to page 100 and select another answer.

3.6d. **Incorrect.** This solution is correct for 500 parts in the bin. However, you do not know the number of parts in the bin and cannot assume a number. Therefore, this is a binomial problem based on one of two assumptions—so many parts it is close to independent, or someone is replacing them as fast as you use one.

$$P(1) = \frac{\binom{10}{1}\binom{490}{7}}{\binom{500}{8}} = \frac{\frac{10!}{1!\,9!}\frac{490!}{7!\,483!}}{\frac{500!}{8!\,492!}}$$

$$= \frac{10\,\dfrac{490 * 489 * 488 * 487 * 486 * 485 * 484 * 483!}{7 * 6 * 5 * 4 * 3 * 2 * 1 * 483!}}{\dfrac{500 * 499 * 498 * 497 * 496 * 495 * 494 * 493 * 492!}{8 * 7 * 6 * 5 * 4 * 3 * 2 * 1 * 492!}}$$

$$= \frac{10\,\dfrac{6.49646483\ E18}{5040}}{\dfrac{3.69247042\ E21}{40,320}}$$

$$= \frac{10 * 1.28898112\ E15}{9.15791275\ E16} = \frac{1.28898112\ E16}{2.41150805\ E19}$$

$$= .140750535$$

Go back to page 31 and select another answer.

6.4a. **Correct.** This is the probability of any failures. Draw a picture. Since MTBF is given, use m instead of computing λ. The trick here is realizing that $t = 600$ because there are 300 hours for each system. Use all digits rather than rounding .6666666.

$$P(0) + P(1) + P(2) + \cdots + P(\infty) = 1$$
$$1 - P(0) = P(\text{any})$$
$$P(0) = e^{-t/m} = e^{-600/900} = e^{-.6666666} = .5134171$$
$$1 - P(0) = 1 - .5134171 = .4865829$$

Go to the next problem on page 49.

2.15a. **Incorrect.** This probability is for less than 2 bad resistors. It is almost correct.

$$P(0) = \frac{\binom{10}{0}\binom{990}{6}}{\binom{1000}{6}} = \frac{1 * 1.287912 \text{ E15}}{1.368173 \text{ E15}}$$

$$= .941336984$$

$$P(1) = \frac{\binom{10}{1}\binom{990}{5}}{\binom{1000}{6}} = \frac{10 * 7.845150 \text{ E12}}{1.368173 \text{ E15}}$$

$$= .0573403239$$

$$P(0) + P(1) = .941336984 + .0573403239 = .9986773079$$

Go back to page 22 and select another answer.

2.3d. **Incorrect.** This answer is only one way that you could have a bad resistor. There are two ways.

Go back to page 17 and select another answer.

11.10b. **Incorrect.** This answer is made up. Draw the diagram. Where is C'? Where is D'? Where do the lines intersect?

Go back to page 116 and select another answer

8.5b. **Incorrect.** This is correct for lower bounds. There is one more step for upper bounds.

Go back to page 70 and select another answer.

9.3b. **Correct.** The best way to solve a parallel system is to determine the only way that it could fail and subtract that from 1. To fail, blocks 3 and 4 have to fail.

$$P(f)_3 * P(f)_4 = .3 * .3 = .09$$
$$P(s) = 1 - P(f) = 1 - .09 = .91$$

The network is then

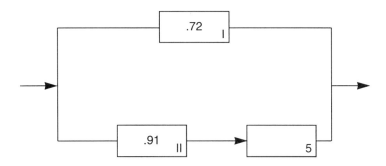

Go to the next problem on page 90.

10.1d. Correct. The simultaneous failure of 4,5,7 prevents the system from working.

Go to the next problem on page 99.

9.11a. Incorrect. This answer is made up. Think about how you can use a previous answer.

Go back to page 92 and select another answer.

11.21b. Incorrect. This answer is made up. Draw the diagram. What is the union of the universe with anything?

Go back to page 118 and select another answer.

14.9b. Maybe. E is running concurrently with f, but it depends on when e actually started between days 12 and 20. For example, if it did not start until day 16, by the time that you were 5 days into f (day 17) there would not be enough time to use the resources from both and finish both jobs. You would have to examine all options at this point before deciding definitely what to do. You may want to borrow resources from two activities.

We hope this guide has helped!

12.2c. Incorrect. $F2$ and $F3$ are on the same level as $F1$ and lead into the fault. They cannot equal anything involving $F1$.

Go back to page 128 and select another answer.

14.2a. **Incorrect.** Select paths by following arrows in the direction in which each points. How did you get from *a* to *e* to *f*? Carefully select each path.

Go back to page 152 and select another answer.

2.10c. **Incorrect.** This solution has two problems. First, you assumed independency. Second, this answer is the probability of no bad resistors. You need to do one more step.

Go back to page 18 and select another answer.

6.1b. **Incorrect.** This cannot possibly be right. Probability is bound by 0 and 1, so no probability can be greater than 1. This is the answer you would get if you set the problem up correctly, but forgot the negative sign in the exponent.

Go back to page 48 and select another answer.

13.7d. **Maybe.** Since this alternative was the alternative of choice under expected value, if you have faith in your numbers, it would be easy to justify its selection. However, if you cannot afford to suffer a -2 loss, you may want to reconsider.

Go to the next answer on page 138.

14.5d. **Incorrect.** At least LF_b is incorrect. You must consider both activities stemming from event 2 (*c* and *d*). To compute LF for LF_b use the earliest length of time from those activities, to allow all activities past that event to remain on schedule. Since event 2 can be complete by 7 (through *c*) or 13 (through *d*), use 7. To compute the correct answer, take LF_c and subtract the duration of *c* from it.

$$LF_b = LF_c - Dur_c = 12 - 5 = 7$$

Go back to page 152 and select another answer.

9.3c. **Incorrect.** This is the probability of both blocks working. It is only one of several ways in which the parallel system will work.

$$P(s)_3 * P(s)_4 = .7 * .7 = .49$$

Go back to page 90 and select another answer.

8.5c. Correct. The degrees of freedom for upper bounds equal $2F + 2$.

Go to the next problem on page 70.

2.15c. Incorrect. This is the probability of exactly 2 bad resistors.

$$P(2) = \frac{\binom{10}{2}\binom{990}{4}}{\binom{1000}{6}} = \frac{45 * 39,782,707,965}{1.368173 \text{ E}15}$$

$$= 1.308475595 \text{ E - }3$$

Go back to page 22 and select another answer.

14.9d. Maybe. G is running concurrently with f, but it depends on when g actually started between days 5 and 15. For example, if it did not start until day 14, by the time that you were 5 days into f (day 17) there would not be enough time to use the resources from both and finish both jobs. You would have to examine all options before deciding definitely what to do. You may want to borrow resources from two activities.

We hope this guide has helped!

7.4a. Correct. First draw a sketch to determine what is desired. Next, determine the z value. Since z is between 1.92 and 1.93, interpolate. Since x is to the right of $\bar{x}$ and you want the probability of $f < x$, add z_{rep} to .5.

$$z = \frac{1500 - 1400}{52} = \frac{100}{52} = 1.923$$

$z_{rep}(1.92) = .47256$

$z_{rep}(1.93) = .47319$

$z_{rep}(1.93) - z_{rep}(1.92) = .47319 - .47256 = .00063$

1.623 is .3 from 1.63

$.3 * .00063 = .000189$

$$z_{rep}(1.623) = .47256 + .000189 = .472749$$
$$P(f < 1500) = .5 + .472749 = .972749$$

Go to the next problem on page 61.

1.13b. Incorrect. This is the solution for permutations, but the notation in the problem is for combinations.

Go back to page 8 and select another answer.

10.8a. Incorrect. At a minimum, there is a minimum cut-set missing.

Go back to page 100 and select another answer.

2.11a. Incorrect. This number is made up. Review the formula and try again.

Go back to page 21 and select another answer.

11.10c. Correct. The diagram shows that the lines intersect everywhere except within C and D. Therefore, find the value for C union D and subtract from 1.

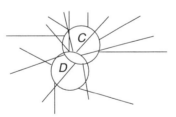

$$P(C' \cap D') = 1 - P(C \cup D) = 1 - (P(C) + P(D) - P(C \cap D))$$
$$1 - (.1 + .05 - .1 * .05) = 1 - (.15 - .005) = .855$$

Go to the next problem on page 116.

14.9c. Incorrect. B has to be over $(LF_b = 3)$ before f begins $(ES = 12)$. Therefore, it would be impossible to borrow resources from e.

Go back to page 153 and select another answer.

11.15c. Incorrect. This answer is made up. Draw the diagram. Where does B' intersect C? Where does this have a union with A?

Go back to page 117 and select another answer.

2.15d. Correct. This is the probability of the video tube catching fire. Since the problem states that the tube will catch fire if 2 resistors are bad, is it not logical to assume it will catch fire with 3 or 4, etc.? Drawing the picture will help solve it.

$$P(0) + P(1) + P(2) + P(3) + P(4) + \ldots + P(10) = 1$$
$$P(\geq 2) = 1 - (P(0) + P(1))$$

$$P(0) = \frac{\binom{10}{0}\binom{990}{6}}{\binom{1000}{6}} = \frac{1 * 1.287912 \text{ E15}}{1.368173 \text{ E15}}$$

$$= .941336984$$

$$P(1) = \frac{\binom{10}{1}\binom{990}{5}}{\binom{1000}{6}} = \frac{10 * 7.845150 \text{ E12}}{1.368173 \text{ E15}}$$

$$= .0573403239$$

$$P(0) + P(1) = .941336984 + .0573403239 = .9986773078$$
$$1 - (P(0) + P(1)) = 1 - .9986773078 = .0013226922$$

This completes the general probability sample problems.

2.8d. Correct. This is the probability of 2 bad resistors in 3 draws. The problem is dependent, and there are 3 ways in which 2 bad resistors could be drawn. Therefore, the addition law must be used.

$$P(b,b,g) = \frac{40}{100} * \frac{39}{99} * \frac{60}{98} = \frac{93,600}{970,200} = .096474954$$

$$P(g,b,b) = \frac{60}{100} * \frac{40}{99} * \frac{39}{98} = \frac{93,600}{970,200} = .096474954$$

$$P(b,g,b) = \frac{40}{100} * \frac{60}{99} * \frac{39}{98} = \frac{93,600}{970,200} = .096474954$$

$$P(b,b,g) + P(g,b,b) + P(b,g,b) = .096474954 + .096474954 + .096474954$$

$$= .28942 = 2.8942 \text{ E-}1$$

Go to the next problem on page 18.

5.1a. Incorrect. This answer is correct for the following setup, which is wrong. Remember your checks to see if the numbers have been correctly inserted into the formula.

$$P(2,4) = \frac{\binom{6}{2}\binom{10}{4}}{\binom{80}{9}} = \frac{15 * 210}{2.3190 \text{ E}11} = \frac{3150}{2.3190 \text{ E}11}$$

$$= 1.35834237 \text{ E-}8$$

Go back to page 39 and select another answer.

3.5b. Correct. This problem is dependent and requires the formula in Chapter 2. There are 8 bad parts from the given information—.02 * 400.

$$P(1) = \frac{\binom{8}{1}\binom{392}{7}}{\binom{400}{8}} = \frac{\dfrac{8!}{1!\,7!}\dfrac{392!}{7!\,385!}}{\dfrac{400!}{8!\,392!}}$$

$$= \frac{8 * \dfrac{392 * 391 * 390 * 389 * 388 * 387 * 386 * 385!}{7 * 6 * 5 * 4 * 3 * 2 * 1 * 385!}}{\dfrac{400 * 399 * 398 * 397 * 396 * 395 * 394 * 393 * 392!}{8 * 7 * 6 * 5 * 4 * 3 * 2 * 1 * 392!}}$$

$$= \frac{8\dfrac{1.34774282\ E18}{5040}}{\dfrac{6.10783814\ E20}{40,320}} = \frac{8 * 2.6740929\ E14}{1.51484081\ E16} = \frac{2.13927432\ E15}{1.51484081\ E19}$$

$$= .141221065$$

Go to the next problem on page 31.

15.2a. Incorrect. This is the probability of all three blocks failing simultaneously. What must every block do for the system to work?

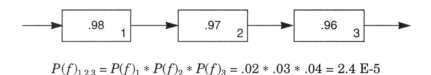

$$P(f)_{1,2,3} = P(f)_1 * P(f)_2 * P(f)_3 = .02 * .03 * .04 = 2.4\ E\text{-}5$$

Go back to page 155 and select another answer.

15.20d. Incorrect. This answer is made up. How should you eat an elephant—one bite at a time. Break this network into small parts and solve each one.

Go back to page 160 and select another answer.

15.14a. Incorrect. This is a binomial problem, but it is the correct answer for 6 or more injuries. A slight correction will provide the correct answer.

$$P(0) + P(1) + P(2) + P(3) + P(4) + P(5) + P(6) + \ldots + P(200) = 1$$

$$P(0) = \binom{200}{0} * (.025)^0 * (.975)^{200} = 1 * 1.0 * .006322994$$

$$= .006322994$$

$$P(1) = \binom{200}{1} * (.025)^1 * (.975)^{199} = 200 * .025 * .0064851$$

$$= .032425638$$

$$P(2) = \binom{200}{2} * (.025)^2 (.975)^{198}$$

$$= \frac{200!}{2! \, 198!} * (.025)^2 * (.975)^{198}$$

$$= 19,900 * .006250 * .0066514$$

$$= .082726948$$

$$P(3) = \binom{200}{3} * (.025)^3 * (.975)^{197}$$

$$= \frac{200!}{3! \, 197!} * (.025)^3 * (.975)^{197}$$

$$= 1.3134 \text{ E6} * .0000156 * .0068220$$

$$= .139999450$$

$$P(4) = \binom{200}{4} * (.025)^4 * (.975)^{196}$$

$$= \frac{200!}{4! \, 196!} * (.025)^4 * (.975)^{196}$$

$$= 64,684,950 * 3.9063 \text{ E-7} * .0069969$$

$$= .176794178$$

$$P(5) = \binom{200}{5} * (.025)^5 * (.975)^{195}$$

$$= \frac{200!}{5! \, 195!} * (.025)^5 * (.975)^{195}$$

$$= 2,535,650,040 * 9.765625 \text{ E-9} * .0071763$$

$$= .177700814$$

$$P(0) + P(1) + P(2) + P(3) + P(4) + P(5) = .615970028$$

$$1 - (P(0) + P(1) + P(2) + P(3) + P(4) + P(5))$$

$$= 1 - .615970028 = .384029972$$

Go back to page 158 and select another answer.

15.25a. Incorrect. This answer is made up. Break this network into small parts and solve each one. Look at blocks 1 and 2 in series. Remember that this is an accident system.

Go back to page 162 and select another answer.

15.18d. Incorrect. This answer is made up. How should you eat an elephant— one bite at a time. Therefore, break this network into small parts and solve each one.

Go back to page 159 and select another answer.

15.15d. Incorrect. Remember, the total of the values in the denominator must equal the numerator, and the total probabilities must equal 1. You must account for all of the flights.

$$P(2,10,20) = \frac{250!}{2!\,10!\,20!} * (.01)^1 * (.04)^{10} * (.1)^{20}$$

$$= \frac{3.2328563\,\text{E}492}{2(3628800)(2.4329020\,\text{E}18)} * .01 * 1.0485760\,\text{E-}14 * 1\,\text{E}20$$

$$= 1.919856\,\text{E}431$$

Go back to page 158 and select another answer.

15.12c. Correct. The panel will malfunction if more than 2 switches are bad. Solve for $P(0) + P(1) + P(2)$ and subtract this from 1 (this is the solution below) or solve for $P(3) + P(4) + P(5)$. Either way results in the same answer.

$$P(0) + P(1) + P(2) + P(3) + P(4) + P(5) = 1$$

$$P(0) = \frac{\binom{15}{0} * \binom{985}{8}}{\binom{1000}{8}} = \frac{1 * 2.1359567\,\text{E}19}{2.4115081\,\text{E}19}$$

$$= .8857348$$

$$P(1) = \frac{\binom{15}{1} * \binom{985}{7}}{\binom{1000}{8}} = \frac{15 * 1.7472038\,\text{E}17}{2.4115081\,\text{E}19} = \frac{2.6208057\,\text{E}18}{2.4115081\,\text{E}19}$$

$$= .1086791$$

$$P(2) = \frac{\binom{15}{2} * \binom{985}{6}}{\binom{1000}{8}} = \frac{105 * 1.2492775 \text{ E}15}{2.4115081 \text{ E}19} = \frac{1.3117414 \text{ E}17}{2.4115081 \text{ E}19}$$

$$= .005439506$$

$$P(0) + P(1) + P(2) = .8857348 + .1086791 + .0054395 = .9998535460$$

$$1 - (P(0) + P(1) + P(2)) = 1 - .9998535460 = .0001465399$$

Go to the next problem on page 157.

15.5a. **Incorrect.** This answer is obtained with a dependent probability formula. The problem appears to be dependent because one is drawing parts out of a bin. It is logical and factual that what one draws on the first selection affects the probability of the second selection. However, to work the problem as a dependent problem, you need all information required for the formula. This answer assumes that there are exactly 1000 fuses in the bin. This assumption is not legitimate. One cannot assume any total. Since the total is unknown, make one of two other assumptions— there are so many fuses in the bin that the probability is so close to being independent that it does not matter (is there really a difference in 1/100,000,000 and 1/99,999,999?), or as a fuse is removed someone else puts one in the bin. Then it would act as an independent problem.

$$P(2,1,3) = \frac{\binom{20}{2}\binom{40}{1}\binom{940}{3}}{\binom{1000}{6}} = \frac{190 * 40 * 137,989,180}{1.3681733 \text{ E}15}$$

$$= \frac{1.0487178 \text{ E}12}{1.3681733 \text{ E}15} = .0007665095$$

Go back to page 155 and select another answer.

15.19d. Correct. This answer is the probability of any accidents based on the data provided. In the multinomial distribution the probability of any acts like a binomial, so you can also solve it that way. The biggest problem here is remembering that the p's must add to 1 and that all of the exposures must be accounted for. This leads to the assumption of the good trips.

$$P(0,0,0,500) = \frac{500!}{0!\,0!\,0!\,500!} * (.0001)^0 * (.0002)^0 * (.008)^0 * (.9899)^{500}$$

$$= 1 * 1 * 1 * 1 * .006246865 = .006246865$$

$$1 - P(0,0,0,500) = 1 - .006246865 = .993753135$$

Go to the next problem on page 160.

15.9d. Correct. This is the probability of a failure given this reliability system. Since there is only one way to fail (all 3 blocks fail at the same time), you could multiply the individual probabilities of failure. However, in a more complicated system, this method may not be as simple. Always solve for system success and subtract from 1.

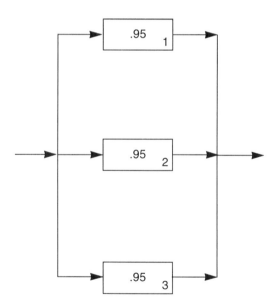

$$P(s)_{sys} = 1 - P(f)_{sys} = 1 - P(f)_1 * P(f)_2 * P(f)_3$$
$$= 1 - .05 * .05 * .05 = 1 - .0001250 = .999875$$
$$P(f)_{sys} = 1 - P(s)_{sys} = 1 - .999875 = .000125$$

Go to the next problem on page 157.

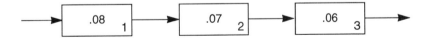

15.6b. Incorrect. This solution is the probability that every block fails. It does not answer any question other than the probability of simultaneous failure of all 3 blocks. What must happen for an accident system in series to have an accident?

$$P(f)_{1,2,3} = P(f)_1 * P(f)_2 * P(f)_3 = .92 * .93 * .94 = .804264$$

Go back to page 155 and select another answer.

15.1a. Incorrect. This answer is obtained by using a wrong a posteriori probability. For an accident on any given trip, the probability is .001 (25/25,000) not .01.

$$P(0) + P(1) + P(2) + P(3) + P(4) + P(5) + \ldots + P(5200) = 1$$
$$P(5) = \binom{5200}{5} * (.01)^5 * (.99)^{5195}$$
$$= \frac{5200!}{5!5195!} * (.01)^5 * (.99)^{5195}$$
$$= 3.162278 \, E16 * 1.0 \, E\text{-}10 * 2.112691 \, E\text{-}23$$
$$= 6.6809 \, E\text{-}17$$

Go back to page 155 and select another answer.

15.20c. Incorrect. This is the probability of failure of the system (no accident). You went one step too far.

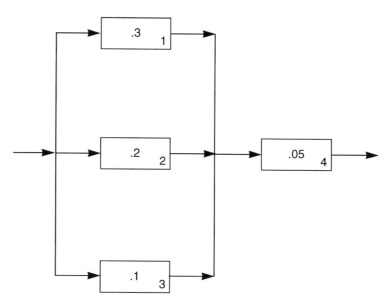

$$P(s)_\text{I} = 1 - P(f)_1 * P(f)_2 * P(f)_3)$$
$$= 1 - (.7 * .8 * .9) = 1 - .504 = .496$$
$$P(\text{acc}) = P(s)_\text{sys} = P(s)_\text{I} * P(s)_4 = .496 * .05 = .02480$$
$$P(f)_\text{sys} = 1 - P(s)_\text{sys} = 1 - .02480 = .9752$$

Go back to page 160 and select another answer.

15.16b. Incorrect. This solution is for exactly 1 bad part.

$$P(0) + P(1) + P(2) + P(3) + P(4) + P(5) - 1$$

$$P(1) = \frac{\dbinom{6}{1} * \dbinom{144}{4}}{\dbinom{150}{5}} = \frac{6 * 17,178,876}{591,600,030} = \frac{103,073,256}{591,600,030}$$

$$= .1742279$$

Go back to page 159 and select another answer.

15.23d. Incorrect. This number is made up. What type of problem is this—dependent or independent? How many cases?

Go back to page 161 and select another answer.

15.13a. Correct. This answer is correct for the probability of no accident in this accident system. The only way an accident will not occur in a parallel accident system is if every block fails at the same time. Always solve for success first and then subtract from 1.

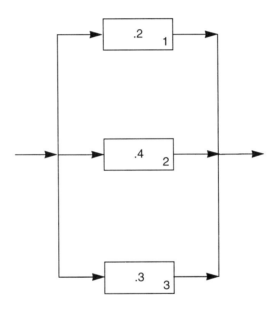

$$P(s)_{sys} = 1 - P(f)_{sys} = 1 - P(f)_1 * P(f)_2 * P(f)_3$$
$$= 1 - .8 * .6 * .7 = 1 - .3360 = .6640$$
$$P(f)_{sys} = P_{no\ acc} = 1 - P(s)_{sys} = 1 - .6640 = .3360$$

Go to the next problem on page 158.

15.8c. **Incorrect.** This is the answer for exactly 3 failures. Although the aircraft will crash if this occurs, it is not the only way.

$$P(0) + P(1) + P(2) + P(3) + P(4) = 1$$

$$P(3) = \binom{4}{3} * (.003)^3 * (.997)^1 = \frac{4!}{3!\,1!} * (.003)^3 * (.997)^1$$

$$= 4 * \; 2.7 \text{ E-8} * \; .997 = 1.07676 \text{ E-7}$$

$$= 1.07676 \text{ E-7}$$

Go back to page 156 and select another answer.

15.25b. **Incorrect.** This is the probability of failure of the system (no accident). You went one step too far.

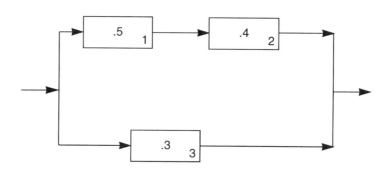

$$P(s)_{\text{I}} = P(s)_1 * P(s)_2 = .5 * .4 = .2000$$
$$P(s)_{\text{sys}} = 1 - (P(f)_{\text{I}} * P(f)_3) = 1 - ((1 - P(s)_{\text{I}}) * P(f)_3)$$
$$= 1 - ((1 - .2) * .7) = 1 - (.8 * .7)$$
$$= 1 - .5600 = .4400$$
$$P(f)_{\text{sys}} = 1 - P(s)_{\text{sys}} = 1 - .4400 = .5600$$

Go back to page 162 and select another answer.

15.11b. Correct

$$P(1,2,4,993) = \frac{1000!}{1!\,2!\,4!\,993!} * (.0001)^1 * (.0009)^2 * (.002)^4 * (.997)^{993}$$

$$= \frac{1000(999)\ldots(994)(993!)}{1!\,2!\,4!\,993!} * .0001 * 8.1\,E\text{-}7 * 1.6\,E\text{-}11 * .0506165$$

$$= \frac{9.7917427\,E20}{48} * .0001 * 8.1\,E\text{-}7 * 1.6\,E\text{-}11 * 0506165$$

$$= .001338184$$

Go to the next problem on page 157.

15.21d. Incorrect. This probability is for no defective resistors. One more step is needed for the solution.

$$P(0) = \binom{4}{0} * (.016)^0 * (.984)^4 = \frac{4!}{0!\,4!} * 1 * .9375197 = .9375197$$

Go back to page 160 and select another answer.

15.2b. Correct. This solution recognizes the fact that in a reliability system in series, every block must function. Since standard convention is to place the probability of success of the block inside the block, simply multiply each block.

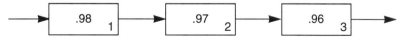

$$P(s)_{\text{sys}} = P(s)_1 * P(s)_2 * P(s)_3 = .98 * .97 * .96 = .912576$$

Go to the next problem on page 155.

15.18b. **Incorrect.** This answer is close in that it is the probability of failure of the system. Recognize that $P(f)$ is $1 - P(s)$ so you could find reliability now by $1 - P(f)$.

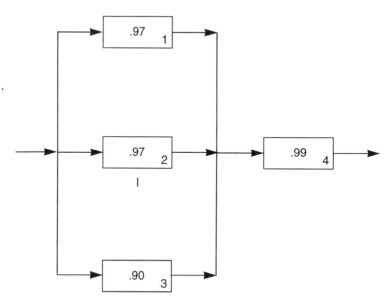

$$P(s)_I = 1 - (P(f)_1 * P(f)_2 * P(f)_3)$$
$$= 1 - (.03 * .03 * .1) = 1 - .00009 = .99991$$
$$P(s)_{sys} = P(s)_I * P(s)_4 = .99991 * .99 = .9899109$$
$$P(f)_{sys} = 1 - P(s)_{sys} = 1 - .9899109 = .0100891$$

Go back to page 159 and select another answer.

15.4d. **Incorrect.** This is the solution for exactly 1 faulty resistor.

$$P(0) + P(1) + P(2) + P(3) + P(4) + P(5) = 1$$

$$P(1) = \frac{\binom{60}{1} * \binom{540}{4}}{\binom{600}{5}} = \frac{60 * 3,503,707,515}{637,262,850,120} = \frac{210,222,450,900}{637,262,850,120}$$

$$= .3298834$$

Go back to page 155 and select another answer.

15.13b. Incorrect. This is the probability of an accident given this accident system in parallel. You went one step too far.

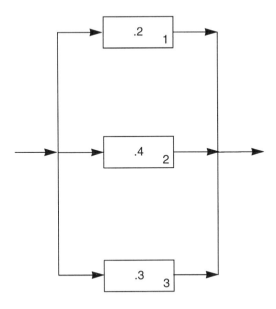

$$P(\text{acc}) \; = \; P(s)_{\text{sys}} = 1 - P(f)_{\text{sys}} = 1 - P(f)_1 * P(f)_2 * P(f)_3$$
$$= \; 1 - (.8 * .6 * .7) = 1 - .3360 = .6640$$

Go back to page 158 and select another answer.

15.10d. Incorrect. This answer is for the failure of only 2 APUs. This is needed to answer the question, but is not correct by itself. What must happen for an aircraft to crash?

$$P(0) \; + \; P(1) \; + \; P(2) \; + \; P(3) \; = \; 1$$
$$P(2) = \binom{3}{2} * (.004)^2 * (.996)^1 \; = \; 3 * 1.6 \; E-5 * .996$$
$$= \; 4.7808 \; E-5$$

Go back to page 157 and select another answer.

15.5b. **Incorrect.** This solution fails to account for the 3 safe fuses that must be in the machine. Remember that the sum of the numbers in the denominator of the first fraction must equal the numerator. Another clue has to do with the sum of the probabilities.

$$P(2, 1, -) = \frac{6!}{2! \, 1!} * (.02)^2 * (.04)^1 = \frac{720}{2 * 1} * .0004 * .04$$

$$= .0057600$$

Go back to page 155 and select another answer.

15.22b. **Correct.** This answer is the probability of component failure due to faulty capacitors. If 2 or more fail, the component fails. This can be determined by $1 - (P(0) + P(1))$. Although this initially looks like a dependent problem, it is not. You cannot assume the total number in the bin. Therefore, it acts like a two-case, independent problem.

$$P(0) + P(1) + P(2) + P(3) + P(4) + P(5) = 1$$

$$P(0) = \binom{5}{0} * (.001)^0 * (.999)^5 = 1 * 1 * .995009990$$

$$= .995009990$$

$$P(1) = \binom{5}{1} * (.001)^1 * (.999)^4 = 5 * .001 * .996005996$$

$$= .004980030$$

$$P(0) + P(1) = .99509990 + .004980030 = .999990019985$$

$$1 - (P(0) + P(1)) = 1 - .999990019985 = 9.980015 \, E\text{-}6$$

Go to the next problem on page 161.

15.15c. **Incorrect.** This solution is made up. It could be derived by making a mistake in multiplying either the .01 or the .1 in the correct formula.

Go back to page 158 and select another answer.

15.25c. **Incorrect.** This answer is made up. Break this network into small parts and solve each one. First, look at blocks 1 and 2 in series. Remember that this is an accident system.

Go back to page 162 and select another answer.

15.3a. **Incorrect.** This is the answer for the probability of exactly 1 fatality based on the given data. What is r?

$$P(0) + P(1) + P(2) + P(3) + P(4) + \ldots + P(1000) = 1$$

$$P(1) = \binom{1000}{1} * (.0002)^1 * (.9998)^{999}$$

$$= \frac{1000!}{1!\,999!} * (.0002)^1 * (.9998)^{999}$$

$$= 1000 * .0002 * .8188782 = .1637756$$

$$= 1.6378 \text{ E-}1$$

Go back to page 155 and select another answer.

15.19c. **Incorrect.** This solution is for no accidents in 500 trips. This answer can be very helpful in the final solution.

$$P(0,0,0,500) = \frac{500!}{0!\,0!\,0!\,500!} * (.0001)^0 * (.002)^0 * (.008)^0 * (.9899)^{500}$$

$$= 1 * 1 * 1 * 1 * .006246865 = .0062469$$

Go back to page 160 and select another answer.

15.12b. **Incorrect.** There are two errors in this solution. First it only provides the probability of exactly 3 faulty switches, whereas the correct answer includes more than just 3 faulty switches. Second, this solution assumes an independent distribution, which this is not. The wrong formula is used to solve this dependent probability.

$$P(0) + P(1) + P(2) + P(3) + \cdots + P(8) = 1$$

$$\binom{8}{3} * (.015)^3\,(.985)^5 = 56 * 3.3750 \text{ E-}6 * .9272165$$

$$P(3) = .0001752439$$

Go back to page 157 and select another answer.

15.17b. Incorrect. This answer might be correct if the problem were dependent. It seems to be dependent because parts are being selected from a bin. It is logical and factual that what one draws on the first selection affects the probability of the second selection. However, to work the problem as a dependent problem you need all the information for the formula. This answer assumes exactly 1000 resistors in the bin. This assumption is not legitimate. Since the total is unknown, you must make one of two assumptions. First, there may be so many resistors in the bin that the probability is actually so close to being independent that it does not matter. The other assumption is that as someone removes a resistor, someone else puts one in the bin. Then it would also act as an independent problem.

$$P(0) + P(1) + P(2) + P(3) + P(4) + P(5) = 1$$

$$P(1,3) = \frac{\binom{20}{1} * \binom{980}{3}}{\binom{1000}{4}} = \frac{\frac{20!}{1!\,19!}\frac{980!}{3!\,977!}}{\frac{1000!}{4!\,996!}} = \frac{20 * 156,385,460}{41,417,124,750}$$

$$= \frac{3,127,079,200}{41,417,124,750} = .0755173$$

$$= 7.5517 \text{ E-}2$$

Go back to page 159 and select another answer.

15.3b. Incorrect. This answer is for 0 fatalities. What must you do now?

$$P(0) + P(1) + P(2) + P(3) + P(4) + \ldots + P(1000) = 1$$

$$P(0) = \binom{1000}{0} * (.0002)^0 * (.9998)^{1000}$$

$$= \frac{1000!}{0!\,1000!} * (.0002)^0 * (.9998)^{1000}$$

$$= 1 * 1 * .8187144 = .8187144$$

$$= 8.1871 \text{ E-}1$$

Go back to page 155 and select another answer.

15.16c. Incorrect. This is the answer for the probability of any parts if the problem had been independent. However, if all data are available and there are parts in a bin, it is usually a dependent problem. This formula is therefore the wrong one to use.

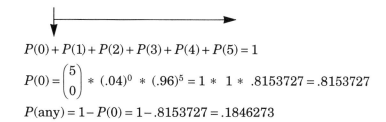

$$P(0) + P(1) + P(2) + P(3) + P(4) + P(5) = 1$$

$$P(0) = \binom{5}{0} * (.04)^0 * (.96)^5 = 1 * 1 * .8153727 = .8153727$$

$$P(\text{any}) = 1 - P(0) = 1 - .8153727 = .1846273$$

Go back to page 159 and select another answer.

15.25d. Correct. Blocks 1 and 2 are in series and are solved first. Then I is in parallel with 3.

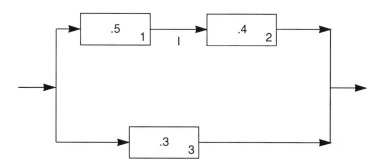

$$P(s)_\text{I} = P(s)_1 * P(s)_2 = .5 * .4 = .2000$$
$$P(s)_\text{sys} = 1 - (P(f)_\text{I} * P(f)_3) = 1 - ((1 - P(s)_\text{I}) * P(f)_3)$$
$$= 1 - ((1 - .2) * .7) = 1 - (.8 * .7)$$
$$= 1 - .5600 = .4400$$

This completes the study guide.

15.5c. **Correct.** This answer is the probability of exactly 2 totally defective fuses and 1 partially defective one. The 3! and the $(.94)^3$ account for the good fuses that are in the machine in question.

$$P(2,1,3) = \frac{6!}{2!\,1!\,3!} * (.02)^2 * (.04)^1 * (.94)^3$$

$$= \frac{720}{2 * 1 * 6} * .0004 * .04 * .8305840$$

$$= .0007973606$$

Go to the next problem on page 155.

15.21c. **Incorrect.** This solution is for any defective resistors if the problem were dependent. However, the critical information of the totals of each type of resistor in the bin is missing. You cannot assume there are 1000 resistors in the bin. Therefore, another assumption must be made.

$$\frac{\binom{1}{0} * \binom{15}{0} * \binom{984}{4}}{\binom{1000}{4}} = \frac{1 * 1 * 38,825,572,626}{41,417,124,750} = .9374280$$

$$P(\text{any}) = 1 - P(0) = 1 - .9374280 = .0625720$$

Go back to page 160 and select another answer.

15.10b. **Incorrect.** This answer is for exactly 1 of the 10 spacecraft failing. This is not the answer for any spacecraft failing. How did you determine p?

$$P(0) + P(1) + P(2) + P(3) + P(4) + \ldots + P(10) = 1$$

$$P(1) = \binom{10}{1} * (4.7872\,\text{E-5})^1 * (.999952128)^9$$

$$= \frac{10!}{1!\,9!} * (4.7872\,\text{E-5})^1 * (.999952128)^9$$

$$= 10 * 4.7872\,\text{E-5} * .999569234$$

$$= .000478514 = 4.7851\,\text{E-4}$$

Go back to page 157 and select another answer.

15.2c. **Incorrect.** This is the probability that all 3 blocks will not fail at the same time. In other words, it is the probability of everything except all 3 blocks failing.

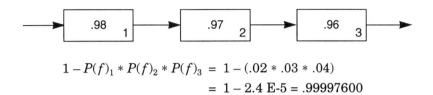

$$1 - P(f)_1 * P(f)_2 * P(f)_3 = 1 - (.02 * .03 * .04)$$
$$= 1 - 2.4 \text{ E-5} = .99997600$$

Go back to page 155 and select another answer.

15.14d. **Incorrect.** This is the probability of 6 or fewer injuries. A minor correction will solve the problem.

$$P(0) + P(1) + P(2) + P(3) + P(4) + P(5) + P(6) + \ldots + P(200) = 1$$

$$P(0) = \binom{200}{0} * (.025)^0 * (.975)^{200} = 1 * 1.0 * .0063230$$
$$= .006322994$$

$$P(1) = \binom{200}{1} * (.025)^1 * (.975)^{199} = 200 * .025 * .0064851$$
$$= .032425638$$

$$P(2) = \binom{200}{2} * (.025)^2 * (.975)^{198} = \frac{200!}{2!\,198!} * (.025)^2 * (.975)^{198}$$
$$= 19,900 * .006250 * .0066514$$
$$= .82726948$$

$$P(3) = \binom{200}{3} * (.025)^3 * (.975)^{197} = \frac{200!}{3!\,197!} * (.025)^3 * (.975)^{197}$$
$$= 1.3134 \text{ E6} * .0000156 * .0068220$$
$$= .139999450$$

$$P(4) = \binom{200}{4} * (.025)^4 * (.975)^{196} = \frac{200!}{4!\,196!} * (.025)^4 * (.975)^{196}$$
$$= 64,684,950 * 3.9063 \text{ E-7} * .0069969$$
$$= .176794178$$

$$P(5) = \binom{200}{5} * (.025)^5 * (.975)^{195} = \frac{200!}{5!\,195!} * (.025)^5 * (.975)^{195}$$

$$= 2,535,650,040 * 9.765625\,\text{E-}9 * .0071763$$

$$= .177700814$$

$$P(6) = \binom{200}{6} * (.025)^6 * (.975)^{194} = \frac{200!}{6!\,194!} * (.025)^6 * (.975)^{194}$$

$$= 82,408,626,300 * 2.441406\,\text{E-}10 * .0073603$$

$$= .1480840$$

$$P(0) + P(1) + P(2) + P(3) + P(4) + P(5) + P(6) = .7640540$$

Go back to page 158 and select another answer.

15.9a. Incorrect. You are close to obtaining the correct answer. Read the question again. This is the probability of reliability for the system. What do you need to do for the probability of failure?

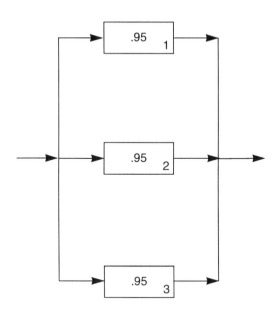

$$P(s)_{\text{sys}} = 1 - P(f)_{\text{sys}} = 1 - P(f)_1 * P(f)_2 * P(f)_3$$
$$= 1 - .05 * .05 * .05 = 1 - .000125 = .9998750$$

Go back to page 156 and select another answer.

15.12a. **Incorrect.** This answers the probability that 3 switches would be defective, but there are other ways for the panel to be defective than to have 3 bad switches.

$$\downarrow$$

$$P(0) + P(1) + P(2) + P(3) + \ldots + P(8) = 1$$

$$\frac{\binom{15}{3} * \binom{985}{5}}{\binom{1000}{8}} = \frac{455 * 7.6486378 \text{ E12}}{2.4115081 \text{ E19}} = \frac{3.4801302 \text{ E15}}{2.4115081 \text{ E19}}$$

$$P(3) = .0001443134$$

Go back to page 157 and select another answer.

15.1b. **Incorrect.** Based on the probability of an accident on any given trip being .001 (25/25000), this answer is the probability of exactly 4 accidents in the 5200 expected trips.

$$\downarrow$$

$$P(0) + P(1) + P(2) + P(3) + P(4) + \ldots + P(5200) = 1$$

$$P(4) = \binom{5200}{4} * (.001)^4 * (.999)^{5196}$$

$$= \frac{5200!}{4!5196!} * (.001)^4 * (.999)^{5196}$$

$$= 3.0429927 \text{ E13} * 1.0 \text{ E-12} * .005524294 = .1681039$$

$$= 1.6810 \text{ E-1}$$

Go back to page 154 and select another answer.

15.8b. Incorrect. This answer is for only 2 engines failing. The aircraft will not crash if only 2 engines fail.

$$P(0) + P(1) + P(2) + P(3) + P(4) = 1$$

$$P(2) = \binom{4}{2} * (.003)^2 * (.997)^2 = \frac{4!}{2!\,2!} * (.003)^2 * (.997)^2$$

$$= 6 * 9.0\,\text{E-}6 * .994009 = .000053676$$

$$= 5.3676\,\text{E-}5$$

Go back to page 156 and select another answer.

15.18c. Correct. To solve this network, solve for the probability of the parallel system first. Then use that reliability times block 4 to obtain system reliability.

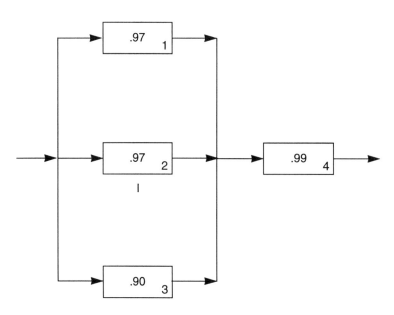

$$P(s)_\text{I} = 1 - (P(f)_1 * P(f)_2 * P(f)_3)$$

$$= 1 - (.03 * .03 * .1) = 1 - .00009 = .99991$$

$$P(s)_\text{sys} = P(s)_\text{I} * P(s)_4 = .99991 * .99 = .9899109$$

Go to the next problem on page 160.

15.16d. **Incorrect.** This solution is for exactly 1 faulty part using an independent formula. Since you are taking parts out of a parts bin, you should assume it is dependent.

$$P(0) \; + \; P(1) \; + \; P(2) \; + \; P(3) \; + \; P(4) \; + \; P(5) \; =$$

$$P(1) \; = \; \binom{5}{1} * (.04)^1 * (.96)^4 \; = \; 5 * .04 * .8493466$$

$$= \; .1698693$$

Go back to page 159 and select another answer.

15.20a. **Correct.** This answer was obtained by solving the individual networks in this system and considering them in series

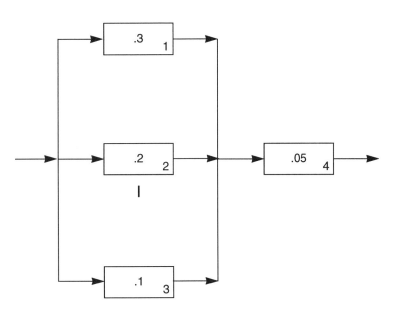

$$P(s)_{\mathrm{I}} \;\; = \; 1 - (P(f)_1 * P(f)_2 * P(f)_3)$$
$$= \; 1 - (.7 * .8 * .9) = 1 - .504 = .496$$
$$P(\mathrm{acc}) \; = \; P(s)_{\mathrm{sys}} = P(s)_{\mathrm{I}} * P(s)_4 = .496 * .05 = .02480$$

Go to the next problem on page 160.

15.22d. Incorrect. This answer is the probability of component failure due to faulty capacitors. If 2 or more fail, the component will fail. The probability is $1 - (P(0) + P(1))$. However, there is rounding error in this solution.

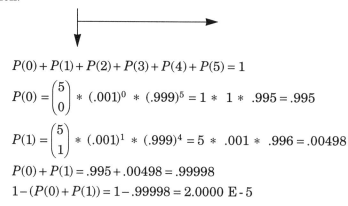

$$P(0) + P(1) + P(2) + P(3) + P(4) + P(5) = 1$$

$$P(0) = \binom{5}{0} * (.001)^0 * (.999)^5 = 1 * 1 * .995 = .995$$

$$P(1) = \binom{5}{1} * (.001)^1 * (.999)^4 = 5 * .001 * .996 = .00498$$

$$P(0) + P(1) = .995 + .00498 = .99998$$

$$1 - (P(0) + P(1)) = 1 - .99998 = 2.0000 \text{ E-}5$$

Go to the next problem on page 161.

15.8d. Incorrect. This answer is for only 1 engine failure. This alone is not close to answering the question. What must happen for the aircraft to crash?

$$P(0) + P(1) + P(2) + P(3) + P(4) = 1$$

$$P(1) = \binom{4}{1} * (.003)^1 * (.997)^3 = \frac{4!}{1!3!} * (.003)^1 * (.997)^3$$

$$= 4 * .003 * .9910270 = .0118923$$

$$= 1.18923 \text{ E-}2$$

Go back to page 156 and select another answer.

15.3d. Correct. This answer is the probability of any fatalities based on the given data. It is computed by subtracting $P(0)$ from 1. The problem is a two-cased, independent one, so use the binomial formula. (Note that it is close to the solution using Poisson 15.3c. This is due to the large number of exposures.)

$$P(0) + P(1) + P(2) + P(3) + P(4) + P(5) + \ldots + P(1000) = 1$$

$$P(0) \quad = \binom{1000}{0} * (.0002)^0 * (.9998)^{1000}$$

$$= \frac{1000!}{0!\,1000!} * (.0002)^0 * (.9998)^{1000}$$

$$= 1 * 1 * .8187144 = .8187144$$

$$P(\text{any}) = 1 - P(0) = .1812856 = 1.8128 \text{ E} - 1$$

Go to the next problem on page 155.

15.7d. Correct. This solution is for exactly 1 faulty relay based on the data provided. The total number of good relays is $500 - 20 = 480$.

$$P(0) + P(1) + P(2) + P(3) + P(4) = 1$$

$$\frac{\binom{20}{1} * \binom{480}{3}}{\binom{500}{4}} = \frac{20 * 18,316,960}{2,573,031,125} = \frac{366,339,200}{2,573,031,125}$$

$$P(1) = .1423765$$

Go back to page 156 and select another answer.

15.19a. Incorrect. This solution is for exactly 1 each of fatalities, major and minor accidents, and 497 safe trips. This is not the solution for any accidents. Remember, any can always be solved by $1 - P(0)$.

$$P(1,1,1,497) = \frac{500!}{1!\,1!\,1!\,497!} * (.0001)^1 * (.002)^1 * (.008)^1 * (.9899)^{497}$$

$$= \frac{500 * 499 * 498 * 497!}{1 * 1 * 1 * 497!} * .0001 * .002 * .008 * .0064400$$

$$= 124,251,000 * .0001 * .002 * .08 * .0064400$$

$$= .001280289$$

Go back to page 160 and select another answer.

15.15b. Incorrect. There is a minor transposition error in this solution. Be careful to match probabilities with their appropriate number of occurrences being predicted.

$$P(-,10,20,218) = \frac{250!}{2!\,10!\,20!\,218!} * (.01)^1 * (.04)^{10} * (.1)^{20} * (.85)^{218}$$

$$= \frac{250!}{2!\,10!\,20!\,218!} * .01 * 1.0485760 \, E\text{-}14 * 1 \, E20 * 4.1051195 \, E\text{-}16$$

$$= \frac{250!}{2(3628800)(2.4329020 \, E18)(4.7402664 \, E416)} * 4.3045298 \, E\text{-}52$$

$$= \frac{3.2628563 \, E492}{8.3699024 \, E441} * 4.3045298 \, E\text{-}52$$

$$= .1662615$$

Go back to page 158 and select another answer.

15.13d. Incorrect. This is the probability of everything except all 3 blocks working simultaneously. This is not close to what you need. Review the rules for parallel systems.

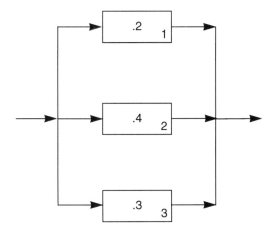

$$1 - P(s)_{1,2,3} = 1 - (P(s)_1 * P(s)_2 * P(s)_3)$$
$$= 1 - (.2 * .3 * .4) = 1 - .02400 = .9760$$

Go back to page 158 and select another answer.

15.10a. **Correct.** First determine that this is a binomial distribution (two-cased, independent). This problem requires two steps. The question can be answered by $1 - P(0)$, where $P(0)$ represents the probability that no spacecraft will fail due to APU failure.

$$P(0) + P(1) + P(2) + P(3) + P(4) + \ldots + P(10) = 1$$
$$P(0) = \binom{10}{0} * (P_{sys})^3 * (1 - P_{sys})^1$$

The problem is that P_{sys} is the probability of spacecraft failure on any given flight, which is unknown. The challenge then is to determine P_{sys}. If 2 or more APUs fail, a spacecraft will fail. P_{sys} can be computed as follows:

$$P(0) + P(1) + P(2) + P(3) = 1$$

This equation can be solved by computing $1 - (P(0) + P(1))$ or $P(2) + P(3)$. Since $P(2) + P(3)$ involves one fewer step, that method is shown. (See 15.4c, page 321, for the other solution.)

$$P(2) = \binom{3}{2} * (.004)^2 * (.996)^1 = 3 * 1.6\,E\text{-}5 * .996$$

$$= .0000478080$$

$$P(3) = \binom{3}{3} * (.004)^3 * (.996)^0 = 1 * 6.4\,E\text{-}8 * 1$$

$$= 6.3744\,E\text{-}8$$

$$P_{sys} = P(2) + P(3) = .000047872$$

Using this, then:

$$P(0) = \binom{10}{0} * (.000047872)^0 * (.999952128)^{10} = 1 * 1 * .999521383$$

$$= .999521383$$

$$1 = P(0) = 1 - .999521383 = .000478617 = 4.7862\,E\text{-}4$$

Go to the next problem on page 157.

15.5d. Incorrect. This solution contains an arithmetic error. The probability of a safe fuse on any draw should be $1 - (.02 + .04) = .94$.

$$P(2, 1, 3) = \frac{6!}{2!\,1!\,3!} * (.02)^2 * (.04)^1 * (.96)^3$$

$$= \frac{720}{2 * 1 * 6} * .0004 * .04 * .8847360$$

$$= .008493466$$

Go back to page 155 and select another answer.

15.2d. Incorrect. This is the probability of failure of this reliability system. Remember that $P(s) + P(f) = 1$, so this is simply $1 - P(s)$.

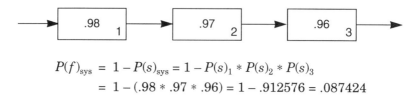

$$P(f)_{sys} = 1 - P(s)_{sys} = 1 - P(s)_1 * P(s)_2 * P(s)_3$$

$$= 1 - (.98 * .97 * .96) = 1 - .912576 = .087424$$

Go back to page 155 and select another answer.

15.23a. Correct. Less than 3 working is the same as saying more than 3 failing. Did you do it this way? There is another way, but it would take more time. The probability of the system working is $P(0) + P(1) + P(2) + P(3)$, but isn't it easier to solve for the only way that it will fail ($P(4)$) and subtract that from 1?

$$P(0) + P(1) + P(2) + P(3) + P(4) = 1$$

$$P(4) = \frac{\binom{10}{4} * \binom{180}{0}}{\binom{190}{4}} = \frac{210 * 1}{52,602,165} = .0000039922$$

$$1 - P(4) = .99999960078$$

Go to the next problem on page 161.

15.11c. Incorrect. This answer is made up. What type of problem is this? What type accidents and their respective probabilities do you know?

Go back to page 157 and select another answer.

15.4a. Incorrect. This answer is obtained from the formula for an independent binomial distribution. This problem is dependent. When one reaches into a parts bin and takes out a part, doesn't the probability of getting a particular part on the next draw change? There is also an error because the probability of a bad part is 10%, which is .1 not .01.

$$P(0) + P(1) + P(2) + P(3) + P(4) + P(5) = 1$$

$$P(0) = \binom{5}{0} * (.01)^0 * (.99)^5 = 1 * 1 * .9509900$$

$$= .95099005$$

Go back to page 155 and select another answer.

15.21b. **Correct.** The information leads one to assume independency since there are no totals for the various types of resistors. Note that since you are only interested in defective and good parts, there are basically only 2 categories, and the multinomial formula can be reduced to the binomial.

$$P(0) = \binom{4}{0} * (.016)^0 * (.984)^4 = \frac{4!}{0!\,4!} * 1 * .9375197 = .9375197$$

$$P(\text{any}) = 1 - P(0) = 1 - .9375197 = .0624803 = 6.2480 \text{ E-2}$$

You could use the multinomial formula and obtain the same results. See the solution for 15.17a on page 312 before going to the next problem on page 161.

15.6d. **Correct.** This is the probability of an accident for this accident system. Regardless of accident or reliability system, for success of the system, systems in series must have every block function as designed.

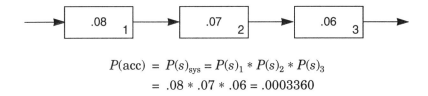

$$P(\text{acc}) = P(s)_{\text{sys}} = P(s)_1 * P(s)_2 * P(s)_3$$
$$= .08 * .07 * .06 = .0003360$$

Go to the next problem on page 156.

15.17d. **Correct.** Although this initially looks like a dependent problem, it is not. You cannot assume the total number in the bin. Therefore, it acts like a two-case, independent problem.

$$P(0) + P(1) + \ldots + P(4) = 1$$

$$P(1) = \binom{4}{1} * (.02)^1 * (.98)^3 = \frac{4!}{1!\,3!} * (.02)^1 * (.98)^3$$

$$= 4 * .02 * .9411920 = .0752954$$

$$= 7.5295 \text{ E-2}$$

Go back to the next problem on page 159.

15.1d. **Incorrect.** This number was made up, and I have no idea how or why you guessed it. Is this problem dependent or independent? What is n? What is r?

Go back to page 154 and select another answer.

15.10c. **Incorrect.** This is the probability of spacecraft failure per exposure. This is a piece of the information needed for the final solution. What must you do now?

$$P(0) + P(1) + P(2) + P(3) = 1$$

$$P(0) = \binom{3}{0} * (.004)^0 * (.996)^3 = 1 * 1 * .9880479$$

$$= .9880479$$

$$P(1) = \binom{3}{1} * (.004)^1 * (.996)^2 = 3 * .004 * .9920160$$

$$= .0119042$$

$$1 - (P(0) + P(1)) = 1 - .9999521 = .000047872$$

Go back to page 157 and select another answer.

15.22a. **Incorrect.** This is one step in determining the answer, but it is not quite there.

$$P(0) + P(1) + P(2) + P(3) + P(4) + P(5) = 1$$

$$P(0) = \binom{5}{0} * (.001)^0 * (.999)^5 = 1 * 1 * .995009990$$

$$= .995009990$$

$$P(1) = \binom{5}{1} * (.001)^1 * (.999)^4 = 5 * .001 * .996005996$$

$$= .004980030$$

$$P(0) + P(1) = .99509990 + .004980030 = .999990019985$$

Go back to page 161 and select another answer.

15.3c. Incorrect. This is the correct answer for the data if the distribution were a Poisson distribution, but there is no indication that it is.

$$P(0) + P(1) + P(2) + P(3) + P(4) + P(5) + \ldots + P(1000) = 1$$

$$\lambda = 1/5000 = .0002 \quad m = 5000/1 = 5000 \quad t = 1000$$

$$P(0) = \frac{(.0002 * 1000)^0 * e^{-.0002(1000)}}{0!} = e^{-.2} = .8187308$$

$$P(\text{any}) = 1 - P(0) = 1 - .8187308 = .1812692$$

Go back to page 155 and select another answer.

15.7c. Incorrect. This answer is the probability of no bad relays.

$$P(0) + P(1) + P(2) + P(3) + P(4) = 1$$

$$\frac{\binom{20}{0} * \binom{480}{4}}{\binom{500}{4}} = \frac{1 * 2,184,297,480}{2,573,031,125}$$

$$P(0) = .8489200$$

Go back to page 156 and select another answer.

15.24d. Incorrect. This answer is made up. How should you eat an elephant—one bite at a time. Break this network into small parts and solve each one. First, look at blocks 1 and 2 in series.

Go back to page 161 and select another answer.

15.17a. Incorrect. This might be correct if the problem were dependent. It seems to be dependent because parts are being selected from a bin. It is logical and factual that what one draws on the first selection affects the probability of the second selection. However, to work the problem as a dependent problem you need all information required for the formula. This answer assumes that there are exactly 100 resistors in the bin. This is not a legitimate assumption. You must make one of two assumptions. First, there may be so many resistors in the bin that the probability is actually so close to being independent that it does not matter. The other assumption might be automatic replacement—as someone removes a resistor, someone else puts one in the bin. Then it would also act as an independent problem.

$$P(0) + P(1) + P(2) + P(3) + P(4) + P(5) = 1$$

$$P(1,3) = \frac{\binom{2}{1} * \binom{98}{3}}{\binom{100}{4}} = \frac{\dfrac{2!}{1!\,1!}\dfrac{98!}{3!\,95!}}{\dfrac{100!}{4!\,96!}} = \frac{2 * 152,096}{3,921,225}$$

$$= \frac{304,192}{3,921,225} = .0775758$$

Go back to page 159 and select another answer.

15.9b. Incorrect. If these blocks were in series, this answer is correct for the probability of reliability and you would only have to subtract from 1 to obtain the correct answer. However, they are not in series. This answer is just the probability that all 3 blocks will function. There are several other ways for the system to function.

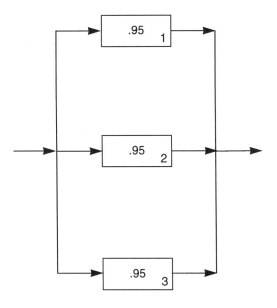

$$P(s)_{1,2,3} = P(s)_1 * P(s)_2 * P(s)_3 = .95 * .95 * .95 = .857375$$

Go back to page 156 and select another answer.

15.4b. Correct. This is the answer for the probability of no faulty resistors based on the information provided. The problem is based on dependent probability, and all of the information necessary is provided. The number of bad and good resistors can be determined by the fact that 10% of them are bad (.10 * 600 = 60 bad and 600 − 60 = 540 good).

$$P(0) + P(1) + P(2) + P(3) + P(4) + P(5) = 1$$

$$\frac{\dbinom{60}{0} * \dbinom{540}{5}}{\dbinom{600}{5}} = \frac{1 * 375,597,445,608}{637,262,850,120}$$

$$P(0) = .5893917$$

Go to the next problem on page 155.

15.13c. Incorrect. This would be the probability of an accident if the system were in series. As it is, this is just the probability that all 3 blocks function, but that is not the only way to obtain an accident in this parallel system.

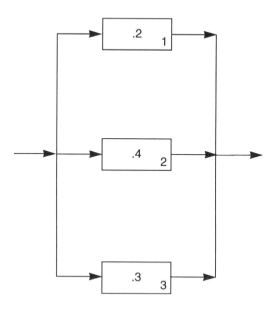

$$P(s)_{1,2,3} = P(s)_1 * P(s)_2 * P(s)_3 = .2 * .3 * .4 = .02400$$

Go back to page 158 and select another answer.

15.23b. **Incorrect.** This answer is the probability that there would be no bad capacitors in the system. This is not the only way that the system will work.

$$P(0) + P(1) + P(2) + P(3) + P(4) = 1$$

$$P(0) = \frac{\binom{10}{0} * \binom{180}{4}}{\binom{190}{4}} = \frac{1 * 42,296,805}{52,602,165} = .8040887$$

Go back to page 161 and select another answer.

15.6c. **Incorrect.** This solution is the probability of everything except the probability of all 3 blocks failing to function properly in the accident system. I think you guessed at this answer. How do you solve an accident system in series?

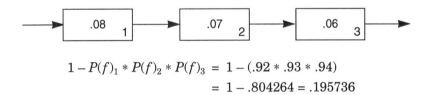

$$1 - P(f)_1 * P(f)_2 * P(f)_3 = 1 - (.92 * .93 * .94)$$
$$= 1 - .804264 = .195736$$

Go back to page 155 and select another answer.

15.20b. **Incorrect.** This answer is made up. This is an accident system. How can you solve for the probability of an accident in the network created by 1,2,3? What would that look like with 4?

Go back to page 160 and select another answer.

15.14b. **Correct.** This answer is the probability of less than 6 injuries based on the data provided.

$$P(0) + P(1) + P(2) + P(3) + P(4) + P(5) + P(6) + \ldots + P(200) = 1$$

$$P(0) = \binom{200}{0} * (.025)^0 * (.975)^{200} = 1 * 1.0 * .0063230$$

$$= .0063229994$$

$$P(1) = \binom{200}{1} * (.025)^1 * (.975)^{199} = 200 * .025 * .0064851$$

$$= .032425638$$

$$P(2) = \binom{200}{2} * (.025)^2 * (.975)^{198}$$

$$= \frac{200!}{2!\,198!} * (.025)^2 * (.975)^{198}$$

$$= 19,900 * .006250 * .0066514$$

$$= .082726948$$

$$P(3) = \binom{200}{3} * (.025)^3 * (.975)^{197}$$

$$= \frac{200!}{3!\,197!} * (.025)^3 * (.975)^{197}$$

$$= 1.3134 \text{ E6} * .0000156 * .0068220$$

$$= .139999450$$

$$P(4) = \binom{200}{4} * (.025)^4 * (.975)^{196}$$

$$= \frac{200!}{4!\,196!} * (.025)^4 * (.975)^{196}$$

$$= 64,684,950 * 3.9063 \text{ E-7} * .0069969$$

$$= .176794178$$

$$P(5) = \binom{200}{5} * (.025)^5 * (.975)^{195}$$

$$= \frac{200!}{5!\,195!} * (.025)^5 * (.975)^{195}$$

$$= 2,535,650,040 * 9.765625 \text{ E-9} * .0071763$$

$$= .177700814$$

$$P(0) + P(1) + P(2) + P(3) + P(4) + P(5) = .6159700$$

Note that the mean number of injuries expected is 5 (.025 * 200). As the probabilities increase from $P(0)$ to $P(5)$, the values increase. What do you expect will happen to the probability for $P(6)$. Why? For part of the answer see 15.5d, page 308.

Go to the next problem on page 220.

15.24c. **Incorrect.** This is the probability of failure of the system. You went one step too far.

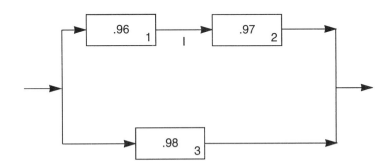

$$P(s)_{\mathrm{I}} = P(s)_1 * P(s)_2 = .96 * .97 = .9312$$
$$P(s)_{\mathrm{sys}} = 1 - (P(f)_{\mathrm{I}} * P(f)_3) = 1 - ((1 - P(s)_{\mathrm{I}}) * P(f)_3)$$
$$= 1 - ((1 - .9312) * .02) = 1 - (.0688 * .02)$$
$$= 1 - .001376 = .998624$$
$$P(f)_{\mathrm{sys}} = 1 - P(s)_{\mathrm{sys}} = 1 - .998624 = .001376$$

Go back to page 161 and select another answer.

15.7b. **Incorrect.** This answer does not account for all of the good relays required to build the part. The total of good and bad relays to build the part must equal the total relays needed. Since we are trying to predict 1 bad relay, the number of good ones must be 3 in order for the total to equal 4.

$$P(0) + P(1) + P(2) + P(3) + P(4) = 1$$

$$\frac{\binom{20}{1} * \binom{480}{2}}{\binom{500}{4}} = \frac{20 * 114{,}960}{2{,}573{,}031{,}125} = \frac{2{,}299{,}200}{2{,}573{,}031{,}125} = .0008935764$$

Go back to page 156 and select another answer.

15.19b.   Incorrect. This is set up to be the solution for no accidents. However, there are arithmetic errors. What does x^0 really equal?

$$P(0,0,0,500) = \frac{500!}{0!\,0!\,0!\,500!} * (.0001)^0 * (.002)^0 * (.008)^0 * (.9899)^{500}$$

$$\neq 1 * .0001 * .002 * .008 * .0062469 = 9.9949841\,E\text{-}12$$

Go back to page 160 and select another answer.

15.15a.   Correct. Although the probability of a safe flight and the total number of safe flights are not explicitly provided, they can be computed based on the other information $(1 - (.01 + .04 + .1) = .85$ and $250 - (2 + 10 + 20) = 218)$.

$$P(2,10,20,218) = \frac{250!}{2!\,10!\,20!\,218!} * (.01)^2 * (.04)^{10} * (.1)^{20} * (.85)^{218}$$

$$= \frac{250!}{2!\,10!\,20!\,218!} * 1\,E\text{-}4 * 1.0485760\,E\text{-}14 * 1\,E20 * 4.1051195\,E\text{-}16$$

$$= \frac{250!}{2(3628800)(2.4329020\,E18)(4.7402664\,E416)} * 4.3045298\,E\text{-}54$$

$$= \frac{3.2328563\,E492}{8.3699024\,E441} * 4.3045298\,E\text{-}54$$

$$= .001662615$$

Go to the next problem on page 159.

15.8a.   Correct. This is the probability of an aircraft crashing due to engine failure. It is derived by computing $P(3) + P(4)$. The problem fits the requirements for a binomial distribution (independent, two-cased). The same answer can be found by subtracting $P(0) + P(1) + P(2)$ from 1.

$$P(0) + P(1) + P(2) + P(3) + P(4) = 1$$

$$P(3) = \binom{4}{3} * (.003)^3 * (.997)^1 = \frac{4!}{3!\,1!} * (.003)^3 * (.997)^1$$

$$= 4 * 2.7\,E\text{-}8 * .997 = 1.076760\,E\text{-}7$$

$$P(4) = \binom{4}{4} * (.003)^4 * (.997)^0 = \frac{4!}{4!\,0!} * (.003)^4 * (.997)^0$$

$$= 1 * 8.1\,E\text{-}11 * 1 = 8.1000\,E\text{-}7$$

$$P(3) + P(4) = 1.07757\,E\text{-}7$$

or

$$P(0) = \binom{4}{0} * (.003)^0 * (.997)^4 = \frac{4!}{0!\,4!} * (.003)^0 * (.997)^4$$

$$= 1 * 1 * .9880539 = .988053892$$

$$P(1) = \binom{4}{1} * (.003)^1 * (.997)^3 = \frac{4!}{1!\,3!} * (.003)^1 * (.997)^3$$

$$= 4 * .003 * .9910270 = .011892237$$

$$P(2) = \binom{4}{2} * (.003)^2 * (.997)^2 = \frac{4!}{2!\,2!} * (.003)^2 * (.997)^2$$

$$= 6 * 9.0\,E\text{-}6 * .9940090 = .0000536765$$

$$1 - (P(0) + P(1) + P(2)) = 1 - .99999989224 = 1.0776\,E\text{-}7$$

Go to the next problem on page 156.

15.11a. Incorrect. The number in the numerator should be the total number of trips. This number is only the number of safe trips. Note that the 993!s cancel. Otherwise most calculators would be unable to compute this answer.

$$P(1,2,4,-) = \frac{993!}{1!\,2!\,4!\,993!} * (.0001)^1 * (.0009)^2 * (.002)^4 * (.997)^{993}$$

$$= \frac{1}{1 * 2 * 24} * .0001 * 8.1\,E\text{-}7 * 1.6\,E\text{-}11 * .05061651$$

$$= 1.3666458\,E\text{-}24$$

Go back to page 157 and select another answer.

15.1c. Correct. Based on the probability of an accident on any trip being .001 (25/25000), this is the probability of exactly 5 accidents in the 5200 expected trips. The problem is binomial because it is two-cased (accident, safe) and independent.

$$P(0) + P(1) + P(2) + P(3) + P(4) + P(5) + \ldots + P(5200) = 1$$

$$P(5) = \binom{5200}{5} * (.001)^5 * (.999)^{5195}$$

$$= \frac{5200!}{5!5195!} * (.001)^5 * (.999)^{5195}$$

$$= 3.1622780 \text{ E16} * 1.0 \text{ E-15} * .00552982 = .1748684$$

$$= 1.7487 \text{ E-1}$$

Go to the next problem on page 154.

15.24b. Correct. Solve blocks 1 and 2 in series and then place block I in parallel with 3.

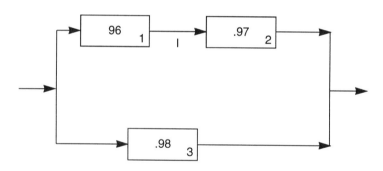

$$P(s)_\text{I} = P(s)_1 * P(s)_2 = .96 * .97 = .9312$$
$$P(s)_\text{sys} = 1 - (P(f)_\text{I} * P(f)_3) = 1 - ((1 - P(s)_\text{I}) * P(f)_3)$$
$$= 1 - ((1 - .9312) * .02) = 1 - (.0688 * .02)$$
$$1 - .0013676 = .998624$$

Go to the next problem on page 162.

15.16a. Correct. The problem is dependent and thus requires this formula. "Any" can always be solved by $1 - P(0)$.

$$P(0) + P(1) + P(2) + P(3) + P(4) + P(5) = 1$$

$$P(0) = \frac{\binom{6}{0} * \binom{144}{5}}{\binom{150}{5}} = \frac{1 * 481,008,528}{591,600,030} = .8130637$$

$$1 - P(0) = .1869363$$

Go to the next problem on page 160.

15.4c. Incorrect. This is the probability of no bad resistors if the probability of bad resistors in the bin had been 1% or .01.

$$P(0) + P(1) + P(2) + P(3) + P(4) + P(5) = 1$$

$$\frac{\binom{6}{0} * \binom{594}{5}}{\binom{600}{5}} = \frac{1 * 605,928,097,368}{637,262,850,120}$$

$$P(0) = .9508292$$

Go back to page 155 and select another answer.

15.6a. Incorrect. This is the probability of no accident for the accident system. If you legitimately obtained this number, you went one step too far. Remember that the probability of failure of the system (no accident in an accident system) is $1 - P(\text{acc})$.

$$P(f)_{\text{sys}} = 1 - P(\text{acc}) = 1 - P(s)_1 * P(s)_2 * P(s)_3$$
$$= 1 - .08 * .07 * .06 = 1 - 3.3600 \text{ E-4} = .9996640$$

Go back to page 155 and select another answer.

15.22c. **Incorrect.** This is one step in determining the answer, but it is not quite there. There is also a rounding error.

$$P(0) + P(1) + P(2) + P(3) + P(4) + P(5) = 1$$

$$P(0) = \binom{5}{0} * (.001)^0 * (.999)^5 = 1 * 1 * .995 = .995$$

$$P(1) = \binom{5}{1} * (.001)^1 * (.999)^4 = 5 * .001 * .996 = .00498$$

$$P(0) + P(1) = .995 + .00498 = .99998$$

Go back to page 161 and select another answer.

15.9c. **Incorrect.** If this were a series system, this answer would be correct. However, since it isn't a series system, this is only the probability of everything except all 3 blocks failing at the same time. All 3 blocks do not have to function for the system to function. Therefore, you cannot subtract $P(s)_{1,2,3}$ from 1 and obtain the correct answer. What must happen in order for the system to fail?

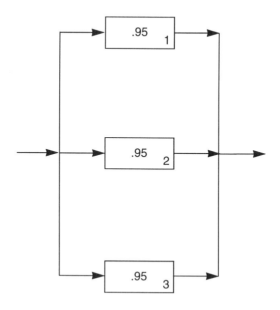

$$1 - P(s)_{1,2,3} = 1 - P(s)_1 * P(s)_2 * P(s)_3$$
$$= 1 - .95 * .95 * .95 = 1 - .857375 = .142625$$

Go back to page 156 and select another answer.

15.21a. **Correct.** The information leads one to assume independency since there are no totals for the various types of resistors.

$$P(0,0,4) = \frac{4!}{0!\,0!\,4!} * (.001)^0 * (.015)^0 * (.984)^4$$
$$= 1 * 1 * 1 * .9375197 = .9375197$$
$$1 - P(0,0,4) = 1 - .9375197 = .0624803$$

Another way to do this problem is a bit shorter in the first step. See the solution for 15.21b on page 310.

15.17c. **Incorrect.** This is the probability of 2 faulty resistors. A minor correction will solve the problem.

$$\downarrow$$

$$P(0) + P(1) + P(2) + P(3) + P(4) = 1$$
$$P(2) = \binom{4}{2} * (.02)^2 * (.98)^2 = \frac{4!}{2!\,2!} * (.02)^2 * (.98)^2$$
$$= 6 * .0004 * .9604000$$
$$= .0023050$$

Go back to page 159 and select another answer.

15.12d. Incorrect. The methodology used is correct, but you rounded too soon. Always leave the entire calculation in the memory of your calculator. Compare this answer with the correct one.

$$P(0) + P(1) + P(2) + P(3) + P(4) + P(5) = 1$$

$$P(0) = \frac{\binom{15}{0} * \binom{985}{8}}{\binom{1000}{8}} = \frac{1 * 2.135956 \text{ E19}}{2.41151 \text{ E19}}$$

$$= .885735$$

$$P(1) = \frac{\binom{15}{1} * \binom{985}{7}}{\binom{1000}{8}} = \frac{15 * 1.74720 \text{ E17}}{2.41151 \text{ E19}} = \frac{2.62080 \text{ E18}}{2.41151 \text{ E19}}$$

$$= .108679$$

$$P(2) = \frac{\binom{15}{2} * \binom{985}{6}}{\binom{1000}{8}} = \frac{105 * 1.24928 \text{ E15}}{2.4115081 \text{ E19}} = \frac{1.311741 \text{ E17}}{2.4115081 \text{ E19}}$$

$$= .0054395$$

$$P(0) + P(1) + P(2) = .885735 + .108679 + .005440 = .999854$$
$$1 - (P(0) + P(1) + P(2)) = 1 - .999854 = .000146$$

Go back to page 157 and select another answer.

15.24a. **Incorrect.** This answer is the probability that everything works. That is not necessary for the system to work. Break the network into its smaller parts to solve correctly.

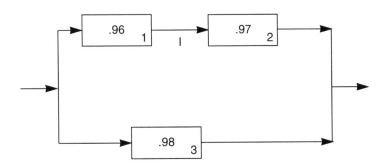

$$P(s)_1 * P(s)_2 * P(s)_3 = .96 * .97 * .98 = .912576$$

Go back to page 161 and select another answer.

15.14c. **Incorrect.** This is the answer for exactly 6 injuries.

$$P(0) + P(1) + P(2) + P(3) + P(4) + P(5) + P(6) + \ldots + P(200) = 1$$

$$P(6) = \binom{200}{6} * (.025)^6 * (.975)^{194}$$

$$= \frac{200!}{6!\,194!} * (.025)^6 * (.975)^{194}$$

$$= 82,408,626,300 * 2.441406\,E\text{-}10 * .0073603$$

$$= .1480840$$

Go back to page 158 and select another answer.

15.18a. Incorrect. This answer was derived by assuming that all of the blocks work. This is nowhere close to what you need to do. First, solve 1,2,3 as a parallel network. Then place that in series with 4.

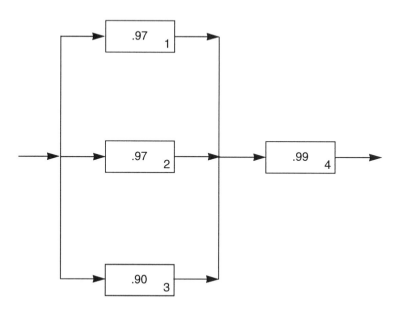

$$P(s)_1 * P(s)_2 * P(s)_3 * P(s)_4 = .97 * .97 * .90 * .99 = .8383419$$

Go back to page 159 and select another answer.

15.7a. Incorrect. This answer reverses the number of good and bad resistors sought. Be sure that the number of bad resistors being predicted is matched with the total number of bad resistors in the bin.

$$P(0) + P(1) + P(2) + P(3) + P(4) = 1$$

$$\frac{\dbinom{20}{3} * \dbinom{480}{1}}{\dbinom{500}{4}} = \frac{1,140 * 480}{2,573,031,125} = \frac{547,200}{2,573,031,125}$$

$$P(3) = .0002126675$$

Go back to page 156 and select another answer.

15.11d. **Incorrect.** This solution has a minor error. You must determine the probability of having a specific number of each type of accident. Some numbers are transposed here

$$P(1,2,4,993) \neq \frac{1000!}{1!\,2!\,4!\,993!} * (.0001)^1 * (.0009)^4 * (.002)^2 * (.997)^{993}$$

$$= \frac{1000(999)\ldots(994)(993!)}{1!\,2!\,4!\,993!} * .0001 * 6.561\,E\text{-}13 * 4\,E\text{-}6 * .0506165$$

$$\neq .000270982$$

Go back to page 157 and select another answer.

15.23c. **Incorrect.** This is the answer for the sum of all of the probabilities except the probability for exactly 3 resistors being bad.

$$P(0) + P(1) + P(2) + P(3) + P(4) = 1$$

$$P(3) = \frac{\binom{10}{3} * \binom{180}{1}}{\binom{190}{4}} = \frac{120 * 180}{52,602,165} = \frac{21,600}{52,602,165} = .000410629$$

$$1 - P(3) = 1 - .000410629 = .999589371$$

Go back to page 161 and select another answer.

Appendix B

TABLES

TABLE B-1. AREAS UNDER THE NORMAL CURVE

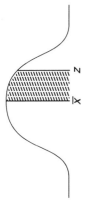

z	0.00	0.01	0.02	0.03	0.04	0.05	0.06	0.07	0.08	0.09
0.0	.00000	.00399	.00798	.01197	.01596	.01995	.02393	.02791	.03189	.03587
0.1	.03984	.04380	.04777	.05173	.05568	.05963	.06357	.06750	.07143	.07535
0.2	.07927	.08317	.04707	.09096	.09484	.09871	.10257	.10642	.11026	.11409
0.3	.11791	.12172	.12551	.12930	.13307	.13682	.14057	.14430	.14802	.15172
0.4	.15541	.15909	.16275	.16639	.17002	.17363	.17723	.18081	.18437	.18792
0.5	.19145	.19496	.19846	.20193	.20539	.20883	.21225	.21565	.21903	.22239
0.6	.22574	.22906	.23236	.23564	.23890	.24214	.24536	.24856	.25174	.25489
0.7	.25803	.26114	.26423	.26730	.27034	.27337	.27637	.27935	.28230	.28523
0.8	.28814	.29103	.29389	.29673	.29955	.30234	.30511	.30785	.31057	.31327
0.9	.31594	.31859	.32123	.32382	.32640	.32895	.33148	.33398	.33646	.33892
1.0	.34135	.34375	.34614	.34850	.35084	.35315	.35544	.35770	.35994	.36215
1.1	.36434	.36651	.36865	.37077	.37287	.37494	.37699	.37901	.38101	.38299
1.2	.38494	.39697	.38878	.39066	.39252	.39436	.39618	.39797	.39974	.40148
1.3	.40321	.40491	.40659	.40825	.40989	.41150	.41309	.41466	.41621	.41774
1.4	.41925	.42074	.42220	.42365	.42507	.42648	.42786	.42923	.43057	.43189
1.5	.43320	.43448	.43575	.43700	.43822	.43943	.44062	.44180	.44295	.44408
1.6	.44520	.44623	.44739	.44845	.44950	.45053	.45154	.45254	.45352	.45449
1.7	.45543	.45637	.45728	.45818	.45907	.45994	.46079	.46163	.46246	.46327
1.8	.46407	.46485	.46562	.46637	.46711	.46784	.46855	.46925	.46994	.47061
1.9	.47128	.47193	.47256	.47319	.47380	.47440	.47499	.47557	.47614	.47670
2.0	.47724	.47778	.47830	.47881	.47932	.47981	.48029	.48076	.48123	.48168
2.1	.48213	.48256	.48299	.48340	.48381	.48421	.48460	.48499	.48536	.48573
2.2	.48609	.48644	.48678	.48712	.48744	.48776	.48808	.48839	.48869	.48898
2.3	.48926	.48954	.48982	.49009	.49035	.49060	.49085	.49110	.49133	.49158
2.4	.49179	.49201	.49223	.49244	.49265	.49285	.49304	.49323	.49342	.49360

TABLE B-1. (cont.)

z	0.00	0.01	0.02	0.03	0.04	0.05	0.06	0.07	0.08	0.09
2.5	.49378	.49395	.49412	.49429	.49445	.49460	.49476	.49491	.49505	.49519
2.6	.49533	.49546	.49559	.49572	.49585	.49597	.49608	.49620	.49631	.49642
2.7	.49652	.49663	.49673	.49683	.49692	.49701	.49710	.49719	.49728	.49736
2.8	.49744	.49752	.49759	.49767	.49774	.49781	.49788	.49794	.49801	.49807
2.9	.49813	.49819	.49824	.49832	.49835	.49841	.49846	.49851	.49856	.49860
3.0	.49865	.49869	.49873	.49877	.49881	.49885	.49889	.49893	.49896	.49900
3.1	.49902	.49906	.49909	.49912	.49915	.49918	.49921	.49923	.49926	.49929
3.2	.49931	.49933	.49936	.49938	.49940	.49942	.49944	.49946	.49948	.49950
3.3	.49951	.49953	.49955	.49956	.49958	.49959	.49961	.49962	.49964	.49965
3.4	.49966	.49967	.49969	.49970	.49971	.49972	.49973	.49974	.49975	.49976
3.5	.49977	.49977	.49978	.49979	.49980	.49981	.49981	.49982	.49983	.49983
3.6	.49984	.49985	.49985	.49986	.49986	.49987	.49987	.49988	.49988	.49989
3.7	.49989	.49990	.49990	.49990	.49991	.49991	.49991	.49992	.49992	.49992
3.8	.49993	.49993	.49993	.49994	.49994	.49994	.49994	.49995	.49995	.49995
3.9	.49995	.49995	.49996	.49996	.49996	.49996	.49996	.49996	.49997	.49997

TABLE B-2. PROBABILITIES FOR A CHI-SQUARE DISTRIBUTION

	Error Rate								
ν	0.75	0.50	0.25	0.20	0.15	0.10	0.05	0.025	0.01
1	0.102	0.455	1.323	1.642	2.072	2.706	3.841	5.024	6.635
2	0.575	1.386	2.773	3.219	3.794	4.605	5.991	7.378	9.210
3	1.213	2.366	4.108	4.642	5.317	6.251	7.815	9.348	11.345
4	1.923	3.357	4.385	5.989	6.745	7.779	9.488	11.143	13.277
5	2.675	4.351	6.626	7.289	8.115	9.236	11.071	12.833	15.086
6	3.455	5.381	7.841	8.558	9.446	10.645	12.592	14.449	16.812
7	4.255	6.346	9.037	9.803	10.746	12.017	14.067	16.013	18.475
8	5.071	7.344	10.219	11.030	12.027	13.362	15.507	17.535	20.090
9	5.899	8.343	11.389	12.242	13.288	14.684	16.919	19.023	21.666
10	6.737	9.342	12.549	13.442	14.534	15.987	18.307	20.483	23.209
11	7.584	10.341	13.701	14.631	15.767	17.275	19.675	21.920	24.725
12	8.438	11.340	14.845	15.812	16.989	18.549	21.026	23.337	26.217
13	9.299	12.340	15.984	16.985	18.202	19.812	22.362	24.736	27.688
14	10.165	13.339	17.117	18.151	19.406	21.064	23.685	26.119	29.141
15	11.037	14.339	18.245	19.311	20.603	22.307	24.996	27.488	30.578
16	11.912	15.339	19.369	20.465	21.793	23.542	26.396	28.845	32.000
17	12.792	16.338	20.489	21.615	22.977	24.769	27.587	30.191	33.409
18	13.675	17.338	21.605	22.760	24.155	25.989	28.869	31.526	34.805
19	14.562	18.338	22.718	23.900	25.329	27.204	30.144	32.852	36.191
20	15.452	19.337	23.828	25.038	26.498	24.412	31.410	34.170	37.566
21	16.344	20.337	24.934	26.171	27.662	29.615	32.671	35.479	38.932
22	17.240	21.337	26.039	27.301	28.822	30.813	33.924	36.781	40.289
23	18.137	22.337	27.141	28.429	29.979	32.007	35.173	38.076	41.638
24	19.037	23.337	28.241	29.553	31.132	33.196	36.415	39.364	42.980
25	19.939	24.337	29.339	30.675	32.282	34.382	37.653	40.647	44.314
26	20.843	25.336	30.435	31.795	33.429	35.563	38.885	41.923	45.642
27	21.749	26.336	31.528	32.912	34.574	36.741	40.113	43.194	46.963
28	22.657	27.336	32.621	34.027	35.715	37.916	41.337	44.461	48.278
29	25.567	28.336	33.711	35.139	36.854	39.088	42.557	45.722	49.588
30	24.478	29.336	34.800	36.250	37.990	40.256	43.773	46.979	50.892
z_α	−.0674	0.000	0.674	0.841	1.036	1.282	1.645	1.960	2.327

Note: For $\nu > 30$: Use formula $c^2 = .5 * (z_\alpha + (2\nu - 1)^{.5})^2$.

BIBLIOGRAPHY

Bar-on, E., and R. Or-Bach, "Program mathematics: A new approach in introducing probability to less able pupils," *International Journal of Mathematical Education in Science and Technology*, 281–297, March/Spring 1988.

Ben-Horim, M., and H. Levy, *Statistics: Decisions and Applications in Business and Economics* (second edition), Random House: New York, 1984.

Freund, John E., *Mathematical Statistics*, Prentice-Hall, Inc.: Englewood Cliffs, New Jersey, 1962.

Gaither, Norman, *Production and Operations Management: A Problem-Solving and Decision-Making Approach*, The Dryden Press: New York, 1987.

Hunstberger, David V., D. James Croft, and Patrick Billingsley, *Statistical Inference for Management and Economics* (second edition), Allyn and Bacon, Inc.: Boston, 1980.

Kavianian, Hamid R., and Charles A. Wentz, Jr., *Occupational and Environmental Safety Engineering and Management*, Van Nostrand Reinhold: New York, 1990.

Kerzner, Harold, *Project Management: A Systems Approach to Planning, Scheduling, and Controlling* (third edition), Van Nostrand Reinhold: New York, 1989.

Mason, Robert D., *Statistical Techniques in Business and Economics* (third edition), Richard D. Irwin, Inc.: Homewood, Illinois, 1974.

Roland, Harold E., and Brian Moriarty, *System Safety Engineering and Management* (second edition), John Wiley and Sons, Inc.: New York, 1990.

Stephenson, Joe, *System Safety 2000: A Practical Guide for Planning, Managing, and Conducting System Safety Programs*, Van Nostrand Reinhold: New York, 1991.

Sutton, Ian S., *Process Reliability and Risk Management*, Van Nostrand Reinhold: New York, 1992.

Vincoli, Jeffrey W., *Basic Guide to System Safety*, Van Nostrand Reinhold: New York, 1993.